··························
Genetics of Asthma and Atopy

Monographs in Allergy
Vol. 33

Series Editors   *L.Å. Hanson*, Göteborg
                    F. Shakib, Nottingham

 Basel · Freiburg · Paris · London · New York ·
New Delhi · Bangkok · Singapore · Tokyo · Sydney

Genetics of Asthma and Atopy

Volume Editor   *I.P. Hall,* Nottingham

20 figures and 5 tables, 1996

KARGER   Basel · Freiburg · Paris · London · New York ·
New Delhi · Bangkok · Singapore · Tokyo · Sydney

·······················

Monographs in Allergy

Library of Congress Cataloging-in-Publication Data
Genetics of asthma and atopy / volume editor, I.P. Hall.
(Monographs in allergy: vol. 33)
Includes bibliographical references and index.
1. Asthma – Genetic aspects.   2. Allergy – Genetic aspects.
I.   Hall, I.P.   II. Series.
[DNLM: 1. Asthma – genetics.   2. Hypersensitivity – genetics.
W1 MO567E v.33 1996 / WF 553 G3283 1996]
RC591.G455   1996
616.2′38042–dc20
ISBN 3–8055–6333–7 (hardcover alk. paper)

Bibliographic Indices. This publication is listed in bibliographic services, including Current Contents® and Index Medicus.

....
Contents

Hall, IP (ed): Genetics of Asthma and Atopy.
Monogr Allergy. Basel, Karger, 1996, vol 33, pp 1–3

....

Introduction

I.P. Hall

Department of Therapeutics, University Hospital, Queen's Medical Centre,
Nottingham, UK

Allergic disease is caused by exposure of genetically predisposed individuals to sensitizing agents. The prevalence of asthma and atopy has been rising in many western countries over the last 20 years, with major social and economic costs. As a consequence, research into the causes of both asthma and allergic disease in general has attracted attention not just from the scientific community but also from the public. Frequent articles in the popular press referring to discovery of 'the asthma gene' have implied that the genetics of allergic diseases are now well understood. The truth, of course, is that we have only begun to understand the mechanisms underlying the genetic predisposition of an individual to develop allergic disease. The aim of this volume is to present reviews in a number of key areas in the field of the genetics of allergic disease. The potential problem with a volume of this kind is that this is a fast-moving area of research and hence reviews may rapidly become dated. To try and counteract this problem, all of the reviews in this volume were contributed during a 6-month period in the second half of 1995.

The genetic analysis of polygenic disease is a complex area for research. Argument still exists about the most appropriate models for complex trait analysis. Genetic studies in allergic diseases have lagged behind studies in other diseases such as Alzheimer's disease, diabetes mellitus and schizophrenia. However, the lessons that have been learnt from attempts to define important genes in these polygenic diseases will be important in genetic studies of allergic diseases.

The first problem that requires an answer is to estimate the contribution of genetic factors in the occurrence of allergic disease and this question is examined in detail in Chapter 2. As well as asthma defined on clinical grounds, three main surrogates have been used in these kinds of study; namely bronchial hyperresponsiveness, serum IgE and the presence or absence of atopy. Whilst some asthmatic individuals will have high IgE levels, atopy and bronchial hyperresponsiveness these traits can and do all exist independently in the general population. The obvious advantage of using IgE levels or bronchial hyperresponsiveness data obtained from inhalation challenge testing for genetic studies is that these measurements provide absolute values, whereas atopy and asthma depend upon clinical definitions which may be subject to variation between studies. From the genetic point of view each of these markers probably represents a different order of complexity. Intuitively, one would imagine that control of IgE levels would be determined by relatively few genes, whereas determination of bronchial responsiveness would be likely to depend upon more factors, and atopy and asthma on an even greater number of factors potentially under genetic control. In general terms, estimates for the contribution of genetic factors to the development of atopy and to the control of IgE responses have led to figures of around a 30–60% contribution, with the rest of the variability being due to environmental factors. The lack of a really good animal model for asthma has significantly hindered research in this area, although some models do exist. These are reviewed in Chapter 3.

One particular difficulty which has not yet been adequately addressed concerns the asthma phenotype. Many studies have used an 'all-or-none' definition, which simple clinical reasoning tells us is clearly a major oversimplification. Asthma is a heterogenous collection of conditions, with a number of overlapping phenotypes, typified at the extremes by the 'extrinsic' asthmatic in whom atopy is a prominent feature and the 'intrinsic' asthmatic in whom atopy is often not present. In addition, analysing asthma as a simple 'all-or-none' disease does not allow for disease severity. Recent work using scoring systems to examine asthma as a semiquantative trait will help here, and may have the advantage that in addition to identifying genes important in causation we may also be able to identify disease-modifying genes.

There are two main aproaches which can be used to determine loci important in the genetics of allergic disease. The first approach which has been used by the majority of studies is that of linkage analysis. A number of studies have used this approach to determine sites of linkage for determinants of allergic disease, often with conflicting results. The early studies suffered from having relatively small numbers of subjects and utilized relatively few markers spread across only limited areas of the human genome. A number of groups are in the process of completing genome-wide searches, and the results of these should be available within the next 2 years. These will provide further evidence for the importance of loci currently under consideration as well as suggesting new loci which may be involved.

Two important areas likely to be involved in control of the allergic response have been identified, namely a region on the long arm of chromosome 5 and a region on the long arm of chromosome 11. The long arm of chromosome 5 contains a number of potential candidate genes for which a biologically plausible role in allergic disease could be proposed including a cluster of cytokine genes (interleukins 4, 5, 9 and 13) and the β_2-adrenoceptor gene. IL-4, IL-5 and IL-13 are important in T_{H2} switching, and hence have been proposed to be central to the development of allergic disease. The gene which has been looked at in most detail on the long arm of chromosome 11 is the gene for the high affinity IgE receptor. Both of these regions and the genes within them are the subject of detailed review in later chapters.

The alternative approach for defining genes important in the development of allergic disease is to perform association studies with genes which potentially could be involved in allergic disease and which contain polymorphic markers. The original studies adopting this approach were performed on the major histocompatibility complex. Chapter 5 contains a review of work examining linkage between HLA haplotypes and allergic disease. More recently the candidate gene approach has been used to examine the possible involvement of a number of genes in the development of allergic disease. Contributions in this volume include reviews of work on tumour necrosis factor (TNF-α) and β_2-adrenoceptor polymorphisms (Chapters 8 and 10 respectively).

Finally it must be remembered that even individuals with a genetic predisposition to allergic disease do not necessarily develop the disease unless they are exposed to the appropriate environmental stimulus. Over the next few years it is likely that a number of major and minor genes which increase the risk of developing allergic disease will be identified, together with disease-modifying genes. The real challenge will then be to examine the complex interplay between genetic and environmental factors in an attempt to understand why a given individual does develop asthma.

Acknowledgements

Putting this volume together has been an enjoyable and educational experience. I have been greatly helped by authors ensuring their contributions have arrived within a defined period. I would like to thank the National Asthma Campaign (UK) for their support of my work. I am also deeply grateful for expert secretarial assistance from Mrs. S. Spiller and Mrs. L. Sabir.

Dr. I.P. Hall, Department of Therapeutics, University Hospital, Queen's Medical Centre, Nottingham NG72UH (UK)

Hall, IP (ed): Genetics of Asthma and Atopy.
Monogr Allergy. Basel, Karger, 1996, vol 33, pp 4–34

....

The Heritability of Allergic Disease

J.C. Dewar, A.P. Wheatley

Department of Therapeutics, University Hospital, Queen's Medical Centre,
Nottingham, UK

Introduction

The atopic diseases are a major cause of morbidity and mortality. Asthma, atopic dermatitis, allergic rhinitis and food allergies together constitute one of the largest groups of clinical disorders requiring medical intervention. The prevalence of these diseases is increasing: that of asthma is approaching 10% in Europe and the USA, whilst as many as 20% of adolescents have hay fever [1]. Atopic dermatitis now affects at least 1 in 10 schoolchildren in Europe [2]. In the United Kingdom alone, there are 3–5 million cases of atopy, leading to as many as 2,000 deaths from asthma each year. The financial burden on the National Health Service therefore is inordinate – the cost of atopic eczema in Britain alone may be as high as GBP 470 million each year [3].

Bearing in mind these statistics, it is vital that we increase our understanding of this group of diseases. However, since the term atopy, which is linguistically derived from the word 'strange', was first coined by Coca and Cooke [4] in 1923, much of the aetiology and pathophysiology of these diseases remains to be clearly delineated, enabling them to retain something of their original mystique.

One of the key issues which requires resolution is that of nature versus nurture: whilst it has long been recognized that there is a familial component to allergy, environmental factors are also important. For example, at least 40% of the population is atopic but only one third of these will at some stage develop allergic disease [5] – this suggests that an external factor is required to trigger the development of allergic disease in a genetically predisposed individual. Thus, allergy is a complex condition which is likely to arise from the interaction of environmental factors, genetic predisposition and individual end-organ specificity, and to attempt to divide the aetiology of allergy into discrete genetic and environmental factors would clearly be an over-simplification.

Genetic factors are undoubtedly important in the aetiology of allergic disease: the familial aggregation of asthma was first observed by Sennertus in 1650, and subsequently, the pioneering work of Cooke and Van der Veer [6] published in 1916 confirmed the familial nature of atopic disease. Numerous subsequent family studies have reported a positive family history in 40–80% of allergic individuals [7]. In the general population the risk of allergy is believed to be about 20%, but this risk increases to 50% if one parent is allergic and further to 66% if both parents are allergic [8]. However, whilst it is clear that there is a hereditary component to allergic disease, this has not been explained in precise genetic terms. Several modes of inheritance have been suggested, including single dominant gene with partial penetrance [6, 8] and single recessive gene with partial penetrance [9, 10]. More recently, maternal inheritance has been implicated via chromosome 11q13 [11]. However, it is highly unlikely that a single gene alone is responsible for the development and transmission of allergic disease, and rather allergy is likely to be a polygenic disorder resulting from the interaction of multiple genes [12]. This inevitably means that the study of genetics in this area is much more complex than that of monogenic disorders.

This chapter will attempt to assess the contribution of genetic factors to the development of allergic disease. Part of the uncertainty in determining the mode, extent and significance of inheritance in allergic disease is due to difficulties in research: the lack of a clear definition of atopy and asthma; selection bias, and failure to control and account for interaction with environmental factors has led to inconsistency in results and confusion over their interpretation. These and other methodological problems will be reviewed in the first section. The evaluation of this complex disease phenotype will then be simplified by individually considering the sub-phenotypes of immunoglobulin E (IgE), atopy, bronchial hyperresponsiveness (BHR), and asthma respectively, assuming that each sub-phenotype may be produced by a smaller number of the multiple genes, which then interact to produce the more complex disease phenotype. Evidence from population studies, family studies and twin studies will be presented and evaluated for each. Finally, the role of the environment and its relative contribution to the development of allergic disease will be assessed.

Methodological Considerations

Research into the genetics of allergic disease has been considerably hampered by methodological inconsistencies. Many of these will be highlighted in the relevant sections, but a few of importance will now be discussed in more detail.

Table 1. Definitions for asthma, atopy, BHR and IgE

Asthma

The American Thoracic Society currently defines asthma as variable airflow obstruction documented by variation in the FEV_1 or the peak expiratory flow rate at >20% spontaneously or >15% after medication, and also requires the presence of chronic symptoms such as wheeze [13].

Atopy

Atopy is generally regarded as a disorder of IgE responses to common environmental allergens, and is detected by either elevated total serum IgE levels, antigen-specific IgE responses, or positive skin prick testing. There is no clear definition of atopy, and some authors require the presence of all three of the above parameters, whilst others are less stringent and consider a positive finding for only one to indicate atopy. Similarly, there is much variation in opinion over what constitutes a positive skin prick test, and whilst a weal size of 4 mm is commonly taken to imply atopy, accepted values range between 3 and 12 mm.

Bronchial hyperresponsiveness

By convention an individual is said to demonstrate BHR if the FEV_1 falls by 20% from the baseline at any standard concentration of inhaled histamine or methacholine [15]. There is a continuous distribution of BHR in the general population, and therefore the point at which hyperresponsiveness is defined is arbitrary – this is reflected in the wide range of values utilized in research [16]. However, a PD_{20} value of 8 µmol is commonly accepted as a cut-off point for asthmatics and nonasthmatics.

IgE

The frequency of atopic disease increases when serum IgE levels exceed 100 U/ml and is very high at serum IgE levels of 700 U/ml. Thus, a serum IgE level of >100 U/ml is usually taken to imply the presence of atopy [17].

Definitions

At present there are no clear or consistent definitions of asthma and atopy which are generally utilized. Similarly, whilst serum IgE and bronchial hyperreactivity are continuous variables, the parameters for normal and abnormal values are arbitrary, and consequently vary between studies. Serum IgE and BHR also show temporal variation, and are affected by many other factors such as age and sex, and this is often not accounted for in the statistical analysis.

Accepted criteria do exist for the definition of asthma [13] but there is little differentiation between the many different clinical entities such as intrinsic and extrinsic, occupational, and late and early onset asthma, and this oversimplification of a complex disease inevitably causes difficulties. The issue is further complicated by the fact that symptomatic and detectable asthma is only present intermittently in some individuals, and thus some true asthmatics will be categorized as nonasthmatics and vice versa. The difficulty in defining asthma is well illustrated when the accuracy of questionnaires used for this purpose is considered: of subjects defined as asthmatic on the basis of a questionnaire, only 44.8% were later clinically proven to be asthmatic [14].

Because of the laxity in definition for atopy and athma, a subjective quality is introduced into research in this area, thus leading to difficulty reproducing, validating, and generalizing results. For clarity, the most commonly used definitions for asthma, atopy, elevated IgE, and BHR are shown in table 1.

Selection Bias and Sample Size

Many studies use small samples, thus reducing their power, and increasing the confidence intervals of results. Studies often recruit subjects from hospital respiratory clinics, or select atopic individuals from the general population – this leads to ascertainment bias, thus limiting the validity and interpretation of results, producing samples which are not representative of the population at large.

Statistical Analysis

There is great variation in the statistical methods employed in the study of genetic epidemiology, causing difficulty in comparing and contrasting results. For instance, many studies express their results as an 'H' value – this provides an estimate of the contribution of heredity (versus environment) to the variance of the measured traits of allergic disease. When H equals 1, the variation in the measured trait is due completely to heredity. As H approaches zero, the trait is presumed to be more environmentally determined. However, the magnitude of heritability calculated depends in part on the statistical analysis employed, and varies accordingly [18].

Many studies fail to perform regression and multivariate analysis, therefore failing to control for the influence of confounding factors, such as smoking and age. The power of the study is often not calculated, and the confidence intervals and significance values not stated.

Family Studies

Families of allergic individuals share a similar genetic background and familial environment, and it is therefore difficult to quantitate the relative contribution of each in the development of allergic disease. The analysis of large numbers of families together may dilute the effects of single gene inheritance operating in specific families. Similarly, the inclusion of such a family with multiple affected individuals will inevitably skew results. The sibling-pair method of analysis is therefore useful: it is a nonparametric analysis which does not make any assumptions about the mode of inheritance, and thus generates more robust results.

Twin Studies

These provide a realm of information in the assessment of the contribution of genetic factors in allergic disease, and are extremely useful. However, they are based on the assumption that twins have a shared environment and therefore that environmental factors are constant – this is obviously an oversimplification.

The process of twinning per se may influence the expression of a trait, and so the results of twin studies may not be applicable to populations of nontwins [19]. Dizygotic twins have been shown to be less interested in the process of twinship than monozygotic twins, and thus more likely to enter studies because of exposure or symptomatology, leading to a degree of selection bias [20]. Also, female twins are more likely to live together than male twin pairs, especially at an older age, and are therefore more likely to be exposed to similar environmental factors [20].

The analysis of twin studies is often performed by casewise and probandwise concordance: this method compares monozygotic twin pairs who are positively concordant or discordant for a disease to similar groups of dizygotic twins, and evaluates the conditional probability that the disease will occur in a second twin when one twin is already affected. However, this form of analysis is only valid if a specific method and its statistical assumptions are clearly defined [21].

Therefore, although twin studies are invaluable, because of these problems, the heritability estimates from twin data should be considered as an upper estimate only [22].

Summary

There are many problems inherent to the study of the genetics of allergy, and these should be considered during the following discourse, when the contribution of genetic factors to the development of allergic disease will be assessed in detail.

Immunoglobulin E

IgE was first discovered in 1967 by Johansson and Bennich [23], who identified an atypical myeloma protein and the corresponding protein in normal serum. This protein was subsequently found to be elevated in the sera of patients with asthma and hay fever and was identified as IgE. Later studies then proved that a raised level of IgE is a major empirical risk factor for the development of allergic disease [24].

The measurement of IgE by the radioallergosorbent assay (RAST) is a relatively simple and sensitive technique which has been shown to have 96% concordance with allergy provocation tests [25]. Work by Spitz et al. [17]

illustrated that the observed frequency of atopic disease increases when serum IgE levels exceed 100 U/ml and is very high at serum IgE levels of 700 U/ml. Thus, a serum IgE level of 100 U/ml is usually taken to imply the presence of atopy. When evaluating total serum IgE concentration it must be remembered that it shows a continuous skewed distribution in the general population and is also influenced by age [26] and smoking [27]. Thus, reliable studies of the relationship between IgE and atopy need to account for these factors in their statistical analysis. Finally, it should be noted that because serum IgE is a continuous variable, it is likely that at a low serum IgE a multitude of other factors may be relevant to the occurrence of allergic disease, whereas a higher serum IgE level may be a more specific risk factor.

The expression of IgE responses and concomitant allergies is probably a function of several genetic factors: Marsh et al. [28] postulate that there are three levels of control in allergic disease: firstly, the genetic regulation of basal serum immunoglobulin production; secondly, control of the immune responses to specific antigens determined by immune regulatory genes linked to the major histocompatibility complex, and thirdly, the existence of an HLA-associated hyperresponsive state controlled by an additional genetic factor responsible for the overall expression of immune response. Thus an individual possesses a predetermined genetic ability to produce exuberant amounts of IgE but specific IgE responses will be triggered by specific environmental allergens.

There have been numerous studies evaluating the inheritance of IgE and the most important will now be outlined.

Family Studies

The first family study performed in this area was by Marsh et al. [29] in 1974: 28 families were investigated, and total IgE was measured by the double antibody radioimmunoassay method. By postulating a cut-off point between low and high IgE phenotype at an arbitrary value of 95 U/ml, a recessive model of Mendelian inheritance was seen at high levels of basal serum IgE, although complex segregation analysis was not performed. In order to investigate this further, Gerrard and Horne [30] studied IgE levels in the parents and children of 80 families. Their results were consistent with low levels of IgE being determined by two dominant genes, the absence of one or another permitting high levels to occur. In 1978, Gerrard et al. [31] went on to analyse total IgE levels in 173 small nuclear families (including 28 families with a high prevalence of atopic disease in the sample), by both path analysis and segregational analysis. This form of statistical analysis allows the detection of major single genes in the presence of polygenic heritability and shared sibling environment. The skewed distribution of IgE can sometimes simulate a major locus, and this was duly accounted for. Their findings gave a value for the heritability of basal serum IgE of 0.425, and supported the hypothesis of a regulatory locus for IgE, which was again thought to be inherited in a recessive manner at high levels of IgE, in concordance with the previous work by Marsh et al. [29]. However, they also identified a significant component of polygenic heritability.

The next family study of total serum IgE levels was the first one to utilize large pedigrees rather than nuclear families [32]. Using a mixed model analysis, Blumenthal et al. [33] found no evidence of significant polygenic component. The estimate of heritability obtained was 49.5% and this was in agreement with the previously outlined study of Gerrard et al. [31]. However, ascertainment bias was a problem in this study because the pedigrees were selected via multiple members with ragweed allergy. Subsequently, further work suggested that in some of these families a dominant major gene for ragweed allergy may have been segregating, thus confounding results. In another study by the same group of seven large pedigree families, the inheritance of IgE in five families was consistent with a recessive mode of inheritance whilst in the other two a dominant mode seemed to be operating [7].

A study of Pennsylvanian Amish pedigrees is the only one to examine the genetics of total IgE levels in an inbred population, and here a codominant model of inheritance was postulated [34]. In a further study of five Mormon pedigrees, the heritability of total IgE levels was estimated as being approximately 60%, but no evidence of a major gene was found [35]. Since the Mormon population is not inbred, these results were thought to be more applicable to the general Caucasian population than those of the Amish study.

Thus it can be seen that the results generated by these studies were conflicting, causing increasing confusion over the issue of IgE genetics. In an attempt to resolve this, Meyers et al. [36] conducted a large study of 278 individuals, comprising 42 nuclear families. Each family was selected for the study through a parent who worked at a local electric company, and were chosen for their large size rather than the presence of allergic disease, thus reducing ascertainment bias. Segregation analysis showed that 36% of total phenotypic variation in log concentration of IgE could be attributed to genetic factors. These were thought to be equally divided between Mendelian and polygenic components, and at high levels of serum IgE autosomal recessive inheritance was believed to operate.

The results of this well-conducted study by Meyers et al. [36] clearly illustrate that there is no simple solution to the obviously complex genetics of IgE. It seems likely that the inheritance of serum IgE is largely under polygenic control, although there may be a major single gene operating in some families. The inheritance of basal IgE will be more clearly defined by the linking of IgE regulatory genes to other markers by linkage analysis, or by the candidate gene approach. There has been much recent work in this area, and in 1989 atopic serum IgE responses were linked to chromosome 11q13 in a study of nuclear families [37], and a dominant mode of inheritance was postulated [38]. More recently, maternal inheritance has been implicated with the transmission of atopy at the chromosome 11q13 locus detectable only through the maternal line [11]. This area of research will be explored in more detail in later chapters.

Twin Studies

Although the problems with twin studies have been outlined previously, these provide a unique opportunity to study the expression of allergic disease in genetically identical individuals. In general, twin studies estimate that serum IgE levels have a heritability in the range of 50–84% [39]. The total serum IgE levels in nearly all twins studies have revealed greater concordance for monozygous twins than for dizygous twins but great variations in estimates for heritability exist. The results of three twin studies will now be presented and discussed:

The first study to be described was performed by Basaral et al. [40] in 1974. A sample of 54 monozygous and 39 dizygous adult twins consisting of American military veterans aged between 45 and 55 years, and 10 monozygous and 13 dizygous twin children aged between 2 and 9.5 years of age, selected from the Crippled Children Division at the University of Oregon Medical School, were assessed. IgE was measured by RAST and was assumed to be normally distributed. In 23 pairs of twins, the IgE was measured on two occasions at 6-monthly intervals – no significant difference was found between the two measures, and thus the temporal variation of IgE was assumed to be negligible. The authors found that both adult and child monozygous twins showed significantly less mean intrapair variance of IgE from that of corresponding groups of dizygous twins ($p < 0.001$). Interestingly, an increased mean intrapair variance of monozygous twin adults relative to the monozygous twin children was seen, and this may indicate that over the long term there is a diversion of IgE levels in individuals with the same genotype. This could be due to the modulation of IgE production by increasing exposure to environmental allergens with age. The authors concluded that 59% of the variance in basal IgE levels in adults and 79% in children could be explained by genetic factors.

In 1984, another large study was performed [18]: 107 pairs of twins were selected via an asthma family study, twin clubs, public schools and local media announcements, with an age range of 6–31 years. Again, significantly less mean intrapair variance of serum IgE was found in monozygous twins relative to dizygous twins ($r = 0.82$ and 0.52 respectively, $p = 0.01$ using a modified Fisher Z transformation). However, when the total variance in log serum IgE was compared between monozygous and dizygous twins, no significant difference was found between the two groups (log IgE 2.20 and 1.94 respectively). Despite this, a heritability estimate of 0.61 was obtained, suggesting a genetic influence on the determination of total serum IgE levels. However, there were several problems inherent to this study: a questionnaire was used to diagnose asthma and atopy, which has been shown to be an extremely unreliable method; no age adjustment was made for serum IgE values, and finally, the prevalence of allergic disease was higher in the study sample than in the general population, reflecting a degree of ascertainment bias.

Finally, a recent American study on monozygotic and dizygotic twins reared together and apart was well conducted and subjected to thorough statistical analysis [41]. Subjects were obtained from the University of Minnesota Twin Registry, and the Finnish National Public Health Institute, without the knowledge of pre-existing atopic disease, so decreasing ascertainment bias. The sample consisted of 68 twins reared apart and 121 twins reared together. Basal serum IgE was measured by the double antibody radioimmunoassay (PRIST) [42]. Specific serum IgE was evaluated by RAST and was corrected for age and sex by regression analysis. A biometric model of analysis was used, allowing variance for genetic factors, nonfamilial and familial environmental factors to be assessed and controlled for. The authors found that whether reared together or apart, monozygotic twins had significantly greater correlation coefficients for total serum IgE levels. They estimated that approximately 50% of the variation in total IgE levels could be attributed to genetic factors. Also, monozygotic twins showed 71% concordance for specific IgE responses to common allergens. The effects of shared familial environments were thought to be negligible because concordance between monozygotic and dizygotic twins reared apart was greater than that for twins reared together. Interestingly, no significant difference in the measurement of specific IgE between monozygotic and dizygotic twins reared apart and together was found, indicating that sensitivity to a particular antigen may be influenced more by environmental than genetic factors. This finding is supported by other studies which have found that the overall regulation of total IgE is largely heritable, but that specific IgE levels are influenced by mainly environmental factors [43, 44].

Thus it seems that genetic make-up determines the ability to produce excessive IgE, but the exposure to environmental allergens may trigger the production of specific IgE, so determining the development of allergic disease. This helps in part to explain why raised basal serum IgE is not necessarily synonymous with symptomatic allergy.

Migrant Studies
The comparison of indigenous populations with immigrant populations allows the genetic influence of allergic diseases to be ascertained, providing that possible cultural differences are accounted for in the statistical analysis.

High serum IgE levels have been noted most commonly in populations that are not of European descent [45]. This difference may simply be the result of an increased exposure to parasites, and in particular the presence of helminth infection. However, in both Ethiopian [46] and Filipino populations [47], serum basal IgE levels of healthy individuals are elevated relative to IgE levels in populations of Euopean descent. Interestingly, elevated serum IgE levels have been observed among infants of Chinese descent in the USA [48], and among children of Chinese descent in England [49]. Both these racial groups have also been found to have a high incidence of atopic disease relative to that seen in their parents' country of origin. One theory which attempts to explain this difference is that helminthiasis, which is endemic to these countries, reduces the likelihood of developing allergic disease.

In order to investigate the effect of racial differences on the production of IgE further, a study of serum IgE levels in Filipino children in the USA was carried out [45]. The study population was small, and consisted of 27 patients of bilateral Filipino descent, aged between 5 and 17 years of age, and 24 Caucasian control patients of a similar age distribution. Filipinos were screened for helminthiasis, and serum IgE was measured by RAST. Interestingly, the serum IgE levels and the frequency of atopy were much higher in the Filipino group than the Caucasian group. Although cultural effects may account in part for these results, the group concluded that genetic differences between the two groups were of most importance. Traits that are selectively neutral may occur in isolated populations simply as a result of genetic drift, and this mechanism may be responsible for the elevated IgE levels observed in Filipinos. However, the ability to produce high levels of serum IgE may confer a selective advantage against helminth disease in the indigenous Filipino population, and it could be speculated that, in the absence of helminth infection in immigrant populations, a high basal serum IgE manifests itself in atopic disease.

Summary

It is apparent that the production of serum basal IgE is under genetic control. Although the exact mode of inheritance remains unclear, it is most likely to be polygenic, and approximately 50% of the variation in basal IgE may be explained by genetic factors. The recent linkage of IgE with markers on chromosome 11q13 [11] and also chromosome 5q31-33 [50] will hopefully lead to the genetics of IgE being more clearly delineated in the near future.

Atopy

Atopy is generally regarded as a disorder of IgE responses to common environmental allergens. The existence of atopy may or may not give rise to the development of asthma, hay fever and eczema. Atopy is detected by either elevated total serum IgE levels, antigen-specific IgE responses, or skin prick testing. Definitions of atopy vary, and some authors require only the presence of increased serum IgE to indicate atopy, whereas others are more stringent.

Atopy and raised IgE do not always coexist – up to a third of individuals with atopic eczema, for example, have normal IgE concentrations. This finding can be partially explained when one considers that IgE production in response to specific allergens is a transient phenomenon which then initiates a more prolonged immune response involving mast cells, eosinophils and macrophages, subsequently leading to the development of symptoms and disease.

There are only a few family and twin studies investigating the inheritance of atopy per se, as it is generally divided into its separate components of serum IgE, specific IgE and skin tests, and seldom studied as an entire entity.

Family Studies

One study has shown a 0–20% risk of atopy if neither parent had atopy, 30–50% risk with one atopic parent and a 60–100% with both parents atopic [51]. Other studies indicate a stronger maternal than paternal influence on atopy: for instance, when the mother alone was atopic, 48% of the offspring were also atopic and 37% were asthmatic, whereas if the father alone was atopic, atopy was seen in 33% and asthma in 25% of the children [52].

Twin Studies

In the twin study, outlined in the previous section on IgE by Hopp et al. [18], serum IgE, complete prick and intradermal skin testing with antigens that were common to the Midwest (including house dust, danders of cats and dogs, mixed antigens of ragweed, grass mould, trees and western weeds) were measured. Individual skin tests responses were scored and a total ISTS (intradermal skin test score) was given as the sum of the individual antigen scores. They also assessed the presence of atopy by means of a questionnaire and the results were analysed by casewise concordance. No significant difference between monozygotic and dizygotic twins was found for a history of atopy. However, the study does not outline what it defines as atopy, and indeed it is hard to know how one can define atopy merely from a history. There was no statistical difference between monozygotic and dizygotic twins for individual skin prick tests for mixed ragweed antigen, cat dander and house dust allergen. Only the mixed tree antigen ISTS showed a significant difference (p < 0.01). Total ISTS scores had a correlation coefficient of 0.82 for monozygotic twins and 0.46 in dizygotic twins and this produced a heritibility factor of 0.72.

Hanson et al. [41] measured single dose prick skin testing to *Ambrosia artemisiifolia, Aspergillus fumigatus, Alternaria teni, Phleum pratense* and *Lo-*

lium perenne. A positive skin test response was defined either as a mean weal diameter of 5 mm or greater for the titration intradermal testing, using a titre of 10^{-3} or less, or as a concentration of 1 in 500 or less for prick testing. They found no significant difference in concordance between monozygotic and dizygotic twins. However, there was a trend for greater concordance seen in the monozygotic twin group but sample sizes were too modest to achieve statistical significance.

In a study of five pairs of monozygotic twins raised apart and four pairs raised together, skin test results showed three of the five sets were positive for at least one antigen among both twins, and 75% of the skin tests exhibited concordance between sets of twins. Three of the four sets of twins raised together had the same allergic history and 70% of the skin tests were concordant. However, the small sample size of this study limits the power of these findings [53]. Another study chose to study skin prick testing in association with HLA haplotype. Forty-five pairs of monozygotic and 28 pairs of dizygotic twins aged between 6 and 30 years were studied. All subjects were tested for ten common allergens by the prick and intradermal methods. Gross diameters of the skin tests were measured and individual allergen and total skin test scores were determined. For monozygotic twins the correlation coefficient for the total skin test score was 0.85, and for dizygotic twins, 0.6 ($p < 0.01$ for both types of twins). Similar results showing significant correlation coefficient in monozygotic but not dizygotic twins were found with moulds, tree pollens and dog and cat dander. No significant differences were found with histamine, house dust, feathers, horse dander or grass pollen [54].

Summary
Whilst it is clear that genetic factors influence the development of atopy, the exact mode of inheritance will only be delineated by studies which analyse the total serum IgE, specific IgE, and skin tests individually, and then attempt to assess the additive effects of these in determining the development of allergic disease. In order to do this, abnormal values for total serum IgE, specific IgE, and skin prick tests require standardization, as do their measurements, and a consistent definition of atopy needs to be agreed.

Bronchial Hyperresponsiveness

The relationship between asthma and airway responsiveness was first described in 1921 by Alexander and Paddock [55]. They observed that asthmatics develop bronchoconstriction in response to cholinergic stimuli more readily than nonasthmatics. Subsequently, bronchoconstriction was demonstrated by giving intravenous histamine to asthmatic subjects [56], and later it was noted that this response did not occur in nonasthmatic subjects, even with larger doses of histamine [57]. Asthmatic subjects are now known to display airway hyperresponsiveness to a large variety of substances including pharmacological agonists such as histamine, methacholine and serotonin, and also the stimulatory prostaglandins. Physical stimuli such as exercise [58] and the inhalation of cold air [59] also induce bronchoconstriction. For this reason, the term nonspecific airway hyperresponsiveness has been used to distinguish between specific airway responses to sensitizing agents such as allergens.

By convention an individual is said to demonstrate BHR if the FEV_1 falls by 20% from the baseline at any standard concentrations of inhaled histamine or methacholine [15]. Methacholine is generally regarded as a more reliable measure because the use of histamine is associated with the phenomenon of tachyphylaxis [60], but the airway response to both substances has been found to be similar in a number of studies [61–63]. Airway responsiveness is measured in a standardized method, and the techniques of Yan et al. [64], Cockcroft et al. [65] and Chai et al. [66] are the most commonly used.

BHR is now recognized as a characteristic and cardinal feature of asthma [67] and is included in the definition utilized by the American Thoracic Society [13]. However, it should be noted that individuals with BHR do not necessarily have asthma or vice versa, and indeed considerable biological variability can be demonstrated for this type of response [68]. There is a continuous distribution of nonspecific airway responsiveness in the general population, with asthmatics in one tail of this distribution, and therefore the point at which hyperresponsiveness is defined is an arbitrary one [68]. In 1985, Hargreave et al. [16] stated that methacholine or histamine bronchial challenge tests are a very sensitive indicator of the presence or absence of current asthma, and are more sensitive than the history, diurnal variation of peak expiratory flow rate and exercise tests. However, the results of a large random sample (n = 2,053) of children in Auckland, New Zealand, seem to contradict this: approximately 50% of children with BHR to histamine had never been diagnosed as being asthmatic, whereas 50% of those diagnosed as being asthmatic did not exhibit BHR. Here, the authors concluded that the degree of overlap in BHR between asthmatics and nonasthmatics was too wide to allow useful clinical prediction [69]. Other studies have shown similar results [70, 71], and the predictive value of BHR ranges between 26 and 47%. However, in polarized groups, BHR has a greater predictive value (79%) and a sensitivity of 87% [72].

The reasons for finding normal bronchial responsiveness in subjects with current asthma are unclear, but there are several possible explanations: (1) atopic status has been shown to influence the level of BHR, independently of the presence of asthma [73], (2) BHR changes over time: Josephs et al. [74]

found a variable relationship of BHR to symptoms and peak expiratory flow rate over time, in repeated studies of the same subjects. In a study of a cohort of children followed from birth and tested with methacholine at 9, 11 and 13 years of age, only 29% of children with wheezing at every interview had BHR on all three occasions, and a further 30% never exhibited BHR [75]. (3) BHR varies with seasonal allergen exposure [76], industrial matter exposure [77], smoking [78], viral respiratory tract infections [79], and age [15].

Thus the relationship between asthma and BHR is not a simple one, and the two are clearly not synonymous. BHR cannot be described as a disease entity resulting from one or two clear-cut aetiological factors, but is likely to be an expression of heterogenous factors, such as polygenic inheritance, which are modified by the environment and other variables such as atopy. The inheritance of bronchial hyperreactivity will now be studied in more detail.

Family Studies

Airway hyperreactivity is frequently found in asymptomatic relatives of asthmatics [80–83] and this is consistent with the notion that BHR represents an inherited sub-phenotype of asthma. However, there are relatively few family studies which investigate this relationship.

One such study investigating the relationship of bronchial hyperreactivity to inhaled carbachol, suggested that BHR is inherited in an autosomal dominant manner [83]. In this study, 50 asthmatics, the healthy parents of 40 asthmatic children, and 70 healthy subjects with no personal or family history of asthma were assessed. Total IgE levels (measured by RAST), and skin prick tests were measured. A bimodal distribution of BHR was seen, raising the issue of whether or not asthmatics represent a hyperreactive population distinct from the normal distribution of airway reactivity, or whether they merely constitute a hyperreactive extreme of a normal distribution pattern. The authors of this study concluded that airway hyperreactivity to carbachol is inherited as an autosomal dominant trait of incomplete penetrance, and is present in 10% of the normal healthy population. They suggest that airway hyperreactivity is a necessary prerequisite, but is not sufficient alone to produce the asthmatic phenotype. This study also provided evidence that BHR and atopy might be under separate genetic control – this would certainly be in keeping with the findings of Sibbald et al. [84], and also recent research linking the high-affinity IgE receptor gene with bronchial hyperreactivity even in the absence of atopy [85]. However, these findings do not concur with the results of a study in the Netherlands of 303 children and grandchildren of probands with asthma. By analysing pairs of siblings, and testing for linkage between BHR and genetic markers on chromosome 5q31–33, it was demonstrated that a trait for an elevated level of serum total IgE, and thus atopy, was coinherited with a trait for BHR [50]. The discovery of the coinheritance and genetic colocalization of the subphenotypes of asthma is exciting, and more will be understood about the complex interactions of asthma, atopy and BHR as new genetic markers and candidate loci are identified.

Twin Studies

Again, there are very few twin studies investigating the inheritance of BHR, and those presented here provide limited information.

In a study of 61 pairs of monozygous and 46 pairs of dizygous twins, selected from an outpatient clinic population, asthma family study and announcements on the local radio, airway hyperresponsiveness was measured by methacholine challenge testing. The response was expressed as the methacholine area, which was calculated by integrating the best fit parabola of the methacholine dose-response curve. The monozygous twins and dizygous twins had intrapair correlation coefficients of 0.67 and 0.43 respectively, and although there was no significant difference in the total variance in methacholine area between the two groups as a whole, a heritability factor of 0.66 was calculated. The authors concluded that there was a significant genetic component to bronchial responsiveness to methacholine [18].

A small study of eight pairs of identical and seven pairs of nonidentical twins, selected so that at least one member of each pair had clinical asthma, or was a first-degree relative of a patient with asthma, assessed the rate of concordance for increased bronchial lability on exercise. They found a significantly higher rate of concordance among the identical twins, both for increased bronchial lability and atopy, and concluded that inheritance was an important factor in both of these characteristics. However, these results must be reviewed with caution in view of the very small sample size, and the rather lax definition of asthma utilized [86].

Summary

It can be seen that whilst there are few studies investigating the inheritance of BHR, those presented here indicate that there is a heritable component. The discovery of new candidate genes and genetic loci will lead to an increased understanding of the complex relationship between BHR, asthma and atopy.

Asthma

Asthma is an increasingly common condition with a prevalence estimated at between 10 and 15%, and a mortality in the order of 1.5–3/100,000 population per year [87]. There is much controversy over what actually constitutes asthma, partly because of its diverse aetiology and the wide range of pathophysiological mechanisms involved. Attempts to define asthma have met with little general agreement and Gross [88] has likened these efforts with attempts to define love! The American Thoracic Society currently defines asthma as variable airflow obstruction documented by variation in the FEV_1 or the peak expiratory flow rate at $> 20\%$ spontaneously or $> 15\%$ after medication, and also requires the presence of chronic symptoms such as wheeze [13]. Whilst this is a useful definition, it does not discriminate between asthma subtypes such as intrinsic and extrinsic, occupational, and aspirin-sensitive asthma. Despite the availability of a clear and simple definition, albeit with limitations, it will be seen that for the purposes of clinical research the diagnosis of asthma is often much less stringent, presenting difficulties in interpreting and generalizing results.

Observations made from lung and bone marrow transplantation have stimulated interest over the issue of genetics versus environment: the transfer of allergen-specific IgE-mediated hypersensitivity has been observed with allogenic bone marrow transplantation [89]. In a report of 4 lung transplant patients, 2 nonasthmatic recipients of asthmatic lungs developed asthma despite prophylactic treatment and immunosuppressive therapy, whilst 2 asthmatic recipients of nonasthmatic lungs did not develop asthma [90]. This finding supports the notion that asthma is a local disease, but the fact that the asthmatic human airway has the capability to remain asthmatic despite as drastic an environmental change as transplantation indicate the importance of inherited factors in determining the behaviour of the asthmatic airway.

Family Studies

The analysis of numerous pedigrees in the early 1900s confirms that there is a heritable component to certain forms of asthma [91, 92]. Since then, family studies have indicated that approximately 26% of the offspring of families ascertained via a parent with asthma have an asthmatic phenotype [93]. The results of three large family studies will now be presented:

In 1980, a study was carried out by Sibbald et al. [84] on a sample of 164 children aged between 1 and 12 years and their families attending a general practice in London. Asthma was diagnosed on the basis of wheezy episodes occurring in response to allergens, exercise or emotion, and the presence of high-pitched wheezing in the chest. Asthmatics were divided into atopic and non-atopic groups on the basis of skin prick testing to allergen. It was concluded that asthma does indeed cluster in families, with a prevalence of 13% in first-degree relatives of asthmatics, whilst that in the relatives of controls being only 4%. These figures agree favourably with those of Leigh and Marley [94], whose family data was collected by a similar method. They found a prevalence of 13.2% in the first-degree relatives of asthmatics and a prevalence of only 1.5% in the relatives of controls. Interestingly, Sibbald et al. [84] concluded that the inheritance of atopic and nonatopic asthma is shared, but that the heritable component of atopic asthma is greater. They also found no strict association between asthma and atopy, suggesting that the two may be inherited independently. This is interesting in the light of recent research which indicates that bronchial hyperreactivity is linked to the inheritance of the high-affinity IgE receptor gene, even in the absence of atopy [85]. Sibbald et al. [84] conclude that clinically different forms of asthma may have a common genetic defect, whose manifestations may be enhanced in the presence of atopy. However, their findings must be interpreted with caution as asthma was diagnosed on the basis of questioning alone and environmental factors were not controlled by regression analysis.

In the East Boston area of Massachusetts, 404 nuclear families were ascertained through a random sample of children, aged between 5 and 9 years, and pulmonary function (FEV_1) was measured on three separate occasions during a 5-year period. The relative contributions of hereditary and environmental factors to individual levels of pulmonary function were assessed by pathwise analysis. Genetic heritability was found to be consistent through time and was estimated to be between 41 and 47%, and was the same for parents and children. Common familial environmental effects on level of pulmonary function explained 1–4% of the variability in children, and 11–28% in adults [95]. This suggests that genetic factors are more influential in childhood, but that environmental factors become more important with increasing age, presumably due to the increasing exposure to allergens which ensues.

Finally, a large prospective study of 1,056 children followed from birth in Christchurch, New Zealand, has generated some interesting results. Family histories were obtained prenatally and the children were followed at 4 months of age and annually to 6 years of age. Other important variables such as home environment, psychosocial factors and family backgrounds were also evaluated. This study confirmed the increased risk of childhood asthma in children of allergic parents. Interestingly, the risk of a child developing asthma was markedly different between the sexes. Boys were much more likely to develop asthma than girls (14.3 vs. 6.3%, p < 0.001). The risk factors for development of atopy between boys and girls also differed: family history of asthma, allergic rhinitis or eczema in either parents or siblings was not significantly related to asthma in girls, the only risk factor being that of eczema in the first year of life [96].

Interestingly, family studies have illustrated that the development of allergy is in part determined by which parent is allergic. Several studies have shown that the risk of developing allergy in children is significantly higher when the mother alone is allergic as compared to the father alone being allergic [97]. For instance, Happle and Schneider [52] evaluated the children of men and women with atopic asthma. In children of atopic fathers, 32 of 98 (33%) were affected by some form of atopic disease compared to 45 of 93 (45%) of the children of atopic mothers. When children with asthma were considered, only 25% had atopic fathers compared to 40% with atopic mothers. These findings were thought to demonstrate the Carter effect: this was first described in relation to pyloric stenosis and the observation that this occurs more frequently in relatives of affected girls than in relatives of affected boys. Carter explained this by proposing that pyloric stenosis was determined by polygenic inheritance and that predisposing genes were equally distributed between the sexes, but the threshold for disease was higher in female than in male subjects. This being the case, female subjects would be less likely to be affected; however, female subjects would have more predisposing genes and thus be more likely to transmit these genes to their offspring [98]. Thus, female asthmatics would be assumed to have a higher threshold for developing asthma and so possess more predisposing genes which they then pass to offspring. This is an interesting theory, but if true one would expect there to be significantly different incidences of asthma between the sexes – there is no evidence to substantiate this. Maternal transmission of allergy has again been implicated in recent research by a team in Oxford. They have linked the transmission of atopic IgE responses to markers on chromosome 11q13, and their work suggests that the sharing of alleles from chromosome 11 by atopic sibling pairs is exclusively from maternal chromosomes [11]. The mechanism by which maternal allergy influences the development of allergy in offspring is unclear: the Carter effect; a maternal effect on fetal and neonatal immune development, and paternal genomic imprinting have all been suggested, but further research is required in order to elucidate the cause of this interesting phenomenon.

Twin Studies

Numerous twin studies have been conducted and four examples will be presented here. One of the largest twin studies was that of Lubs [99] in 1971, who analysed data from the Swedish Twin Registry on 6,996 pairs of adult twins. The study was population based and therefore not subject to the biases induced by the selection of twins by self-referral or from referral populations. The prevalence of asthma in the study was 3.8%, hay fever 14.8% and eczema 2.5%. Concordance rates for asthma, hay fever and eczema were significantly higher in the monozygotic than the dizygotic twins. The concordance rate of 15.4–21.4% in the monozygotic twins was only moderate, and because 79–85% of the monozygotic twins were discordant for these three allergic diseases,

Lubs [99] concluded that whilst the data did support a significant genetic component, the effect of this was small. He then went on to analyse the data from the dizygotic twin pairs in order to determine the relative risk of allergy in a first-degree relative of an allergic individual. Although the risks were again signficant, they were relatively small, ranging from 1.3 to 3.1 times the risk of the general population. He also made the interesting observation that the risk of allergy in first-degree relatives was somewhat disease specific: those relatives of an individual with asthma tended to develop asthma rather than atopic dermatitis or hay fever.

In 1984, Hopp et al. [18] studied 61 pairs of monozygotic and 46 pairs of dizygotic twins, and diagnosed asthma on the basis of a questionnaire. Unlike the previous study, the χ^2 casewise analysis for atopy and asthma history did not demonstrate a significant difference between monozygotic and dizygotic twins (p > 0.05). However, the monozygotic twins were significantly more concordant for methacholine sensitivity in bronchial challenge testing. It may therefore be that the use of a questionnaire in diagnosing asthma has confounded these results, although as previously outlined, the relationship between asthma and BHR is not straightforward.

Another large study of 3,808 twin pairs used the method of path analysis to analyse data on self-reported asthma and hay fever [100]. The cumulative prevalence of asthma or wheezing was 13.2% and 32% for hay fever. There were significant correlations in liability to report disease among twins, and these were higher in monozygotic twins (r = 0.65) than in dizygotic twins (r = 0.25), and in male monozygotic twins (r = 0.75) compared with female monozygotic twins (r = 0.60). Genetic factors were found to be common for both asthma and hay fever, with a correlation in genetic liability to the traits of 0.52 for men and 0.65 for women. The heritability of these diseases was estimated to be 60–70% in this population. Environmental causes of both diseases were also correlated (r = 0.53 for men, r = 0.33 for women).

Finally, in 1991, a population-based study of bronchial asthma in adult twin pairs in Finland was performed. The sample consisted of 27,776 individuals selected from the Finnish twin cohort, of the same gender and born before 1958, with both twins still alive in 1967 [101]. Asthmatic subjects were identified via the hospital admission registry, and the registry for reimbursement of free medication of the social insurance institution. This study was therefore based on a cohort compiled from an entire population, and so decreased the risk of sampling error seen in clinic-based recruitment – this can lead to an increased ascertainment of disease in concordant compared to disconcordant twin pairs. The sample size was also large, so increasing the power of the results obtained. Of the monozygotic twin pairs, 10 were concordant and 138 disconcordant, whereas for the dizygotic pairs 12 were concordant and 343 were disconcordant for asthma. The probandwise concordance rate was 12.7% for monozygotic twins and 6.5% for dizygotic twins. They estimated the heritability factor to be 35.6% for the entire sample, 67.8% for the women and 0% for the men, thus showing a marked gender difference in the inheritance of asthma. This may have been due to chance events effecting distribution of the concordant pairs. The overall heritability estimate for asthma was calculated to be 35.6%.

Population and Migration Studies

Several studies have shown that children of immigrant families born in England have a similar prevalence of asthma to that in European children. However, children of immigrant families, and particularly those of West Indian families, who had been born in the country of origin of their parents have a significantly lower prevalence. This was most marked in relation to families who had immigrated from poor rural areas of Jamaica [102].

A subsequent study of 2,331 children of European, Asian and West Indian parentage attending the Central Birmingham Chest Clinic, found that a higher proportion of children born in England of all races developed asthma within the first 4 years of life than children born abroad (40.2%, p < 0.001). The authors concluded that racial and genetic differences did not contribute to the development of asthma. However, the methods used in this study were not clearly delineated, no definition of asthma was given and the statistical analysis was not outlined [103].

A large study of Chinese, Malay and Indian adults in Singapore drew inconclusive results: 2,868 Chinese, Malay and Indian subjects were studied in five housing estates. Asthma was diagnosed with a questionnaire and statistical analysis used a multivariate approach, so controlling for confounding factors. A cumulative prevalence of asthma of 6.6% in Indians, 6% in Malays and 3% in Chinese was found. The difference between Indians and Malays as a group, and the Chinese was found to be statistically different, although no significance value is stated. The ethnic differences in the prevalence of asthma were still found to be significant after multivariant adjustment for other confounding risk factors such as age and smoking. The authors concluded that these differences may have been due to environmental, cultural or genetic factors [104].

Summary

The family and twin studies presented indicate that between 30 and 50% of the risk of developing asthma may be due to genetic factors. There is some evidence to suggest that the inheritance of asthma is influenced more significantly by the presence of maternal asthma than paternal asthma, and also that the risk factors for the development of asthma may differ between the sexes.

The Environment and Allergy

The recent increase in the prevalence of allergic disease in the last 20 years [105] is too rapid to be explained by changes in the gene pool, and indicates the importance of environmental factors. In order to assess the contribution of genetic factors in the development of allergic disease, the discussion must therefore be balanced by consideration of the role of the environment.

Geographical Variation

Asthma is more frequent in economically developed communities and an increase occurs in the offspring of first-generation migrants from rural to urban environments [105]. For example, asthmatic symptoms in the offspring of Tokelauans who have migrated to New Zealand were twice as great as those reported by those who remained in Tokelau [106]. Similarly, offspring of migrants to Cape Town had an increase in asthma prevalence of 20 times greater than in the migrants' rural community of origin [107].

Cigarette Smoke Exposure

The importance of cigarette smoking in the aetiology of allergic diseases is well illustrated by a study of the populations of the islands of Mogmog and Falalop in the Western Caroline Islands. Ninety percent of the population smoked at least one packet of cigarettes a day and 75% of children under the age of 5 were found to have asthma [108]. Tobacco smoke per se can increase total IgE serum concentrations [109]. However, atopy is less common in smokers, presumably because atopic patients are less likely to take up smoking. Studies of sensitization to occupational allergens, and in particular enzymes, acid anhydrides and platinum salts, have demonstrated that smoking workers have a significantly greater risk of allergy compared to similarly exposed but nonsmoking co-workers [110–112]. Smoking therefore may lower the threshold for specific IgE sensitization.

Maternal smoking seems to have adverse effects on offspring: Maternal cigarette smoking during pregnancy has been associated with increased cord-blood IgE concentrations [113]. Passive exposure to cigarette smoke may increase allergic sensitization in children – the probability of a child having at least one positive skin test is doubled when the mother smokes [114]. Modest increases in total serum IgE concentration in children exposed to cigarette smoke have also been noted [115]. Again, smoking may lower the threshold for sensitization to environmental allergens by influencing immune responsiveness. It may also act directly on airway growth, and indeed babies born to smokers have diminished lung function compared to those born to nonsmokers [105]. Whatever the mechanism, the increase in cigarette smoking of women of child-bearing age has paralleled the increase in asthma and atopic disease in the Western world.

Outdoor Air Pollution

The issue of outdoor air pollution and its contribution to allergic disease, and in particular asthma, is an emotive one. The majority of the general population believes that the two are significantly linked, whereas in fact there seems to be no clear relationship between the frequency and prevalence of asthma and air pollution. However, studies on patients with asthma have shown that air pollutants, and in particular sulphur dioxide, can provoke acute bronchospasm [116]. It may be that pollutants and allergens interact with and modify the actions of each other, and indeed it has been shown that ozone can interact with pollen allergens to reduce the concentration of allergen needed to provoke an athmatic response [117]. One of the most striking pieces of evidence for the implication of airborne pollutants in inducing asthma was seen in the recent outbreaks in Barcelona caused by the unloading of soya beans [118], and in other communities by release of allergens from castor bean factories [119].

Allergen Exposure

Whereas in the developing world allergic responses are often induced by parasites such as *Ascaris lumbricoides*, and as such are necessary and useful physiological responses, in the Western world the more innocuous house dust mite and household pets have been implicated. For example, the introduction of blankets contaminated by house dust mite to the population of New Guinea Highlanders was associated with a striking increase in the prevalence of asthma among a previously little affected community [120]. Positive skin test sensitivity to cats and dogs have been seen in children exposed to family pets under the age of 1 year [97, 121]. These findings suggest that indoor air is an important factor in the development of asthma. The rise in asthma could also be due to improved thermal insulation in homes which, by reducing the number of air changes, causes an increase in the concentration of indoor allergens and pollutants, such as tobacco smoke, oxides of nitrogen from gas cookers, and formaldehyde [105].

Diet

In the 19th century, Salter [122] classified the 'provoking' causes of asthma under three headings: respired irritants; alimentary irritants, and irritants immediately affecting the nervous system. Since then the role of diet in the aetiology of asthma has remained obscure and controversial. No relationship has been found between maternal diet during pregnancy and the development of atopy [123]. However, neonatal diet has been found to influence the development of atopy [105]. There are biologically plausible explanations for the role of both sodium chloride and magnesium in the aetiology of asthma, and there has been much interest in this area. Studies in adults have shown a correlation between male mortality from asthma and table salt consumption [124], an association between airway responsiveness and 24-hour sodium excretion [125] and a reduction in airway responsiveness associated with a reduction in salt intake [126]. However, a large study of 1,702 adults in Nottingham measured 24-hour sodium excretion, BHR to methacholine, atopy (by skin prick testing), and found that a high salt intake did not significantly increase the risk of developing atopy or BHR [127]. The role of dietary magnesium in the aetiology of asthma was investigated in a random population study of 2,633 adults in Nottingham. It was found that an increase in dietary magnesium intake by 100 mg/day was associated with an increase in FEV_1, decreased reporting of wheeze, and decrease in BHR, when age, sex, smoking, height, and atopy were adjusted for [128]. However, the role of magnesium in the treatment of asthma is at present controversial and there is no clear evidence to support the use of inhaled or intravenous magnesium [129, 130].

Breast-Feeding

The role of breast-feeding has been the focus of much controversy, with inconsistency in numerous study findings [131, 132]. However, a recent study following healthy infants during their first year and then through to the age of 17 years has concluded that breast-feeding is prophylactic against atopic disease [133]. Presumably breast-feeding reduces the early exposure to highly allergenic food and inhalation of protein so decreasing the sensitization of genetically predisposed individuals.

Intrauterine Development

The preliminary and controversial work of Barker and co-workers [134] has raised awareness of the importance of intrauterine development of fetuses and a disposition to disease in later life. He has recently shown an association between disproportionate growth of the fetal head in relation to the trunk and limbs, and raised total serum IgE concentrations in adulthood. This association was independent of adult size, social class, smoking, or gestational age at birth. This may simply be a chance association – it is certainly difficult to explain the relationship in causal terms. However, one hypothesis which attempts to explain this phenomenen is that disproportionate fetal growth resulting in a larger head circumference at birth may be associated with impaired thymic development, leading to a diminished production of Th_1 cells which are more sensitive to adverse stimuli [135]. Th_1 cells produce interferon-γ, low levels of which at birth are thought to be a risk factor for the development of atopy [136, 137].

Month of Birth

A number of studies have suggested that the month of birth influences the risk of subsequent allergy [138]. For instance, infants born in Finland between February and April have a maximal risk of sensitization to birch pollen – this period is just prior to the birch pollen season in May. Similarly, the maximal risk of grass and mugwort allergy is found in individuals born between April and May, corresponding to a later pollination season for grass and mugwort [139, 140]. Presumably an intense exposure to an allergen during the first 6–12 months of life, when the immune system is immature, increases the likelihood of developing allergy in later life.

Occupation

The surveillance of work and occupational respiratory disease (Sword Project) suggests that 1,500 new cases of asthma each year in the UK might be attributable to agents inhaled at work. Culprits include isocyanates, flour dust, laboratory animal urine, wood dust and solder flux [105]. Numerous studies have shown that there is an increase in the prevalence of allergic disease in the higher social classes and the reasons for this are unclear – it may simply reflect an increased tendency to report disease with increasing socioeconomic status.

Respiratory Infections

Viral respiratory infections are known to precipitate wheezing in children. Studies by Frick et al. [141] have shown that infants with a bilateral parental history of allergy are significantly more likely to develop allergic symptoms following viral infections. Interestingly, children who have been administered the pertussis vaccine have a marked increase in histamine-induced skin wealing compared with those who have not [142].

It is interesting to note that the risk of developing allergic disease is related to the number of siblings and birth order. A recent Italian study found that skin prick positivity decreased with increasing number of siblings (4 or more versus no siblings, odds ratio 0.37, 95% confidence interval 0.13–1.07, n = 2,226) [143]. Another study conducted in the north of England showed that a subject with no older siblings was 1.71 times more likely to be atopic than a subject with 4 older siblings [144].

Viruses may not be the causal agents, but simply act as triggering factors for the development of asthma during early life. Inflammation of the respiratory tract may increase exposure to airborne allergens at a time when the immune system is immature, and in predisposed individuals an atopic response will ensue.

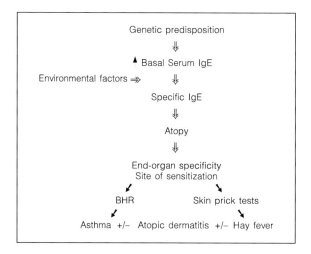

Fig. 1. Interaction of genetic factors, environmental allergens and end-organ sensitivity in the aetiology of allergic disease.

Summary

The above is a brief review of the major environmental factors influencing the development of allergic disease, and it is presumed that the effects of these are summative. An individual's genetic make-up may determine the adequacy or inadequacy of the immune system in regulating responses to these factors, thus determining the development of atopy and subsequent disease.

Conclusion

There is a wealth of research into this interesting but complicated area, and despite methodological difficulties the following conclusions can be drawn:

(1) The aetiology of allergic disease is complex, and likely to result from an interaction of genetic factors, environmental allergens and end-organ sensitivity (fig. 1).

(2) There is clear evidence for the hereditary nature of allergic disease, and the review of the literature indicates that between 30 and 80% of aetiology can be explained by genetic factors. The control of serum IgE appears to be more genetically determined than the expression of atopy, BHR and asthma.

(3) Allergic disease is a polygenic disorder, and several genetic loci may be involved in the overall control and expression of the allergic phenotype. The mode of inheritance will depend on the behaviour of each individual gene – for example, one major gene may be inherited in an autosomal dominant

manner whereas another may be inherited recessively, and so on. The expression and severity of allergic disease will thus be determined by the additive effects of these genes, and the attenuation of their effects by disease-modifying and protective genes.

(4) The variation of severity of allergic disease seen between individuals, and within individuals, may be due to variable gene penetrance and changing environmental factors. The risk of developing allergic disease is probably determined by the separate, and additive risks of each sub-phenotype, and their interaction with the environment. However, the presence of all sub-phenotypes is not required to develop allergic disease and this implies that there is some redundancy in the system, or that there may be other important variables operating which have not yet been recognized.

(5) The following are required if research is to progress: (a) The development of a consistent definition for asthma and atopy and the application of these in genetic studies; (b) the standardization of selection criteria, and (c) agreement on abnormal values for serum and specific IgE, positive skin prick tests and BHR.

(6) Finally, the increasing identification of specific markers and candidate genes, as outlined in later chapters, should significantly enhance the understanding of the inheritance and pathophysiology of this common group of disorders.

References

1   Heywood A: Who's to blame for asthma. Lancet 1995;36:1262.
2   Williams HC: Is the prevalence of atopic dermatitis increasing? Clin Exp Dermatol 1992;17:385–391.
3   Williams HC: Atopic eczema. BMJ 1995;311:1241–1242.
4   Coca AF, Cooke RA: On the classification of the phenomena of hypersensitivity? J Immunol 1923; 8:163–171.
5   Ewan P: Twin to twin atopy (letter). Lancet 1995;345(No. 8963):1508–1509.
6   Cooke RA, Van der Veer A: Human sensitization. J Immunol 1916;16:201 205.
7   Blumenthal MN, Yunis E, Mendell N, Elston RC: Preventative allergy genetics of IgE mediated diseases. J Allergy Clin Immunol 1986;38:962–968.
8   Ratner B, Silverman DE: Allergy: Its distribution and hereditary concept. Ann Allergy 1952;9: 1–20.
9   Weiner AS, Zieve J, Fries JH: The inheritance of allergic disease. Ann Eugen 1936;7:141.
10   Adkinson J: The behaviour of bronchial asthma as an inherited character. Genetics 1920;5:363.
11   Cookson WOCM, Young RP, Sandford AJ, Moffatt MF, Shirakawa T, Sharp PA, Faux JA, Julier C, Le souef PN, Nakumura Y: Maternal inheritance of atopic IgE responsiveness on chromosome 11q. Lancet 1992;340:381–384.
12   Tipps RL: A study of the inheritance of atopic hypersensitivity in man. Am J Hum Genet 1954; 6:3–8.
13   Meneely GR, Renzetti AD, Steel JD, Wyatt JP, Harris HW: Chronic bronchitis, asthma and pulmonary emphysema, definition and classification of chronic bronchitis, asthma and pulmonary emphysema. Am Rev Respir Dis 1962;85:762–768.

14   Cerveri I, Bruschi C, Ricciardi M, Zocchi L, Zoia MC, Rampulla C: Epidemiological diagnosis of asthma: Methodological considerations of prevalence evaluation. Eur J Epidemiol 1987;3:202–205.

15   Townley RG, Bewtra AK, Nair NM, Brodkey FD, Burke BS: Methacholine inhalation challenge studies. J Allergy Clin Immunol 1979;64:569–574.

16   Hargreave FE, Ramsdale EH, Stark PJ, Juniper EF: Advances in the use of inhalation provocation tests in clinical evaluation. Chest 1985;87(suppl):32–35.

17   Spitz E, Gelfand EW, Sheffer AL, Austin KF: Serum IgE in clinical immunology and allergy. J Allergy Clin Immunol 1972;49:337.

18   Hopp RJ, Bewtra AK, Watt GD, Nair NM, Townley RG: The genetic analysis of allergic disease in twins. J Allergy Clin Immunol 1984;73:265–270.

19   Price B: Primary biases in twin studies: A review of prenatal and natal differences producing factors in monozygotic pairs. Am J Hum Genet 1950;2:293–352.

20   Kaprio J, Koskenvuo M, Artimo M, Sarna S, Rantasalo I: Baseline characteristics of the Finnish Twin Registry: Section 1. Material, methods, representativeness and results for variables special to twin studies. Helsinki, Department Public Health Science, 1979, publ M47.

21   Allen G, Harvald B, Shields J: Measures of twin concordance. Acta Genet 1967;17:475.

22   Lewitter FI, Toger IB, McGue M, Tishler PV, Speizer FE: Genetic and environmental determinants of level of pulmonary function. Am J Epidemiol 1984;120:518–529.

23   Johannson SGO, Bennich H: Raised levels of a new Ig class (IgND) in asthma. Lancet 1967;ii:951–953.

24   Berg T, Johannson SGO: IgE concentrations in children with atopic diseases: A clinical study. Int Arch Allergy Appl Immunol 1969;36:219.

25   Gleich GJ, Averbeck AK, Swedlund HA: Measurement of IgE in normal and allergic serum by radioimmunoassay. J Lab Clin Med 1971;77:690.

26   Viller T, Holgate ST: IgE smoking and lung function. Clin Exp Allergy 1994;24:508–510.

27   Burrows B, Hallonen N, Barbee RA, Lebovitz MD: The relationship of serum immunoglobulin E to cigarette smoking. Am Rev Respir Dis 1981;124:523–525.

28   Marsh DG, Hsu SH, Hussain R, Meyers DA, Friedhoff LR, Bias WB: Genetics of human immune response to allergens. J Allergy Clin Immunol 1980;65:322–332.

29   Marsh DG, Bias WB, Morton NE: Genetic control of basal serum IgE levels and its effect on specific reaginic sensitivity. Proc Natl Acad Sci USA 1974;71:3588–3592.

30   Gerrard J, Horne S: Serum IgE levels in parents and children. J Paediatr 1974;85:660–663.

31   Gerrard J, Rao DC, Morton NE: A genetic study of immunoglobulin E. Am J Hum Genet 1978; 30:46–58.

32   Blumenthal MN, Namboodiri KK, Mendell N, Gleich G, Elston RC, Yunis E: Genetic transmission of serum IgE levels. Am J Med Genet 1981;10:219–228.

33   Blumenthal MN, Amos DB, Noreen H: Genetic mapping of Ir locus in man, linkage to second locus. HLA Sci 1974;184:1301–1303.

34   Meyers DA, Bias WB, Marsh DG: Genetic study of total IgE levels in the Amish. Hum Hered 1982;31:15–23.

35   Hasstedt S, Meyers DA, Marsh DG: Inheritance of immunoglobulin E: Genetic model fitting. Am J Med Genet 1983;14:61–66.

36   Meyers DA, Beaty TH, Friedhoff LR, Marsh DG: Inheritance of total serum IgE (basal levels) in man. Am J Hum Genet 1987;41:51–62.

37   Cookson WOCM, Sharp PA, Faux JA, Hopkin JM: Linkage between IgE responses underlying asthma, rhinitis, and chromosome 11q13. Lancet 1989;i:1292–1294.

38   Cookson WOCM, Hopkin JM: Dominant inheritance of atopic immunoglobulin E responsiveness. Lancet 1989;i:86–88.

39   Marsh DG, Blumentahl MN: Immunogenetics of specific immune responses to allergens in twins and families; in Marsh DG, Blumenthal MN (eds): Genetic and Environmental Factors in Clinical Allergy. Minneapolis, University of Minnesota Press, 1990, pp 132–142.

40   Basaral M, Orgel HA, Hamburger RN: Genetics of IgE and allergy serum IgE levels in twins. J Allergy Clin Immunol 1974;54:288–304.

41   Hanson B, McGue M, Roitman-Johnson B, Segal NL, Bouchard TJ Jr, Blumenthal MN: Atopic disease and immunoglobulin E in twins reared apart and together. Am J Hum Genet 1991;48:873–879.

42 Kjellmann IM, Johannson SGO, Roth JA: Serum in healthy children: IgE value quantified by a sandwich technique (PRIST). Clin Allergy 1976;6:51–59.

43 Wutrich B, Baumann E, Fries R, Schnyder U: Total and specific IgE (RAST) in atopic twins. Hum Hered 1983;30:147–154.

44 Bahna SL: Concordance in atopic twins. J Allergy Clin Immunol 1983;71:100.

45 Orgel HA, Lenoir NA, Basaral M: Serum IgG, IgM, IgA and IgE levels and allergy in Filipino children in the United States. J Allergy Clin Immunol 1974;53:213–222.

46 Johansson SGO, Mellbin J, Vahlquist B: Immunoglobulin levels in Ethiopian pre-school children, with special reference to levels of immunoglobulin E. Lancet 1968;i:1118.

47 Rosenberg EB, Whalen GE, Bennich H, Johansson SGO: Increased circulating IgE in a new parasitic disease: Human intestinal capillariasis. N Engl J Med 1970;283:1148.

48 Worth RM: Atopic dermatitis in Chinese infants in Honolulu and San Francisco. Hawaii Med J 1962;22:31.

49 Davis LR, Marten RH, Sarkany I: Atopic excema in European and Negro West Indian infants in London. Br J Dermatol 1961;73:410.

50 Postma DF, Bleecker ER, Amelling PJ, Holroyd KJ, Xu J, Panhuysen CIM, Meyers DA, Levitt RC: Genetic susceptibility to asthma: Bronchial hyperresponsiveness coinherited with a major gene for atopy. N Engl J Med 1995;33:894–900.

51 Kauffman HS, Frick OL: The development of allergy in infants of allergic parents, a prospective study concerning the role of heredity. Ann Allergy 1976;37:410–415.

52 Happel R, Schneider EW: Evidence for the Carter effect in atopy. Int Arch Allergy Appl Immunol 1982;68:90–92.

53 Walsh GG, Roitman-Johnson B, Blumenthal MN: Concordant allergic manifestations among mono-zygotic twins raised apart. J Allergy Clin Immunol 1980;65:215.

54 Townley R, Bewtra A, Watt G, Burke K, Carney K, Nair N: Comparison of allergen skin test responses in monozygous and dizygous twins. J Allergy Clin Immunol 1980;65:214.

55 Alexander HL, Paddock R: Bronchial asthma: Response to pilocarpine and epinephrine. Arch Intern Med 1921;27:184–191.

56 Weiss S, Robb GP: The systemic effects of histamine in man. Arch Intern Med 1932;49:360–396.

57 Curry JJ: The action of histamine on the respiratory tract in normal and asthmatic subjects. J Clin Invest 1946;25:785–791.

58 Anderton RC, Cuff MT, Frith PA, Cockcroft DW, Morse JL, Jones NL, Hargreave FE: Bronchial responsiveness to inhaled histamine and exercise. J Allergy Clin Immunol 1979;63:315–320.

59 O'Byrne PM, Ryan G, Morris M, McCormack D, Jones NL, Morse JL, Hargreave FE: Asthma induced by cold air and its relation to non-specific bronchial responsiveness to methacholine. Am Rev Respir Dis 1982;125:281–285.

60 Manning PJ, Jones GL, O'Byrne PM: Tachyphylaxis to inhaled histamine in asthmatic subjects. J Appl Physiol 1987;63:1572–1577.

61 Townley RG, Ryo UY, Kolotkin BM, Kang B: Bronchial sensitivity to methacholine in current and former asthmatic and allergic rhinitis patients and control subjects. J Allergy Clin Immunol 1975; 56:429.

62 Curry JJ, Lowell FC: Measurement of vital capacity in asthmatic subjects receiving histamine and acetyl-β-methacholine: A clinical study. J Allergy 1948;19:9.

63 Tiffeneau R: Cholinergic and histamine hypersensitivity of lung of the asthmatic. Acta Allergol 1950;5(suppl):137.

64 Yan K, Salome C, Woolcock AJ: Rapid method for the measurement of bronchial responsiveness. Thorax 1983;38:760–765.

65 Cockcroft DW, Killian DN, Mellon JJA, Hargreave FE: Bronchial reactivity to inhaled histamine: A method and a clinical survey. Clin Allergy 1977;7:235–243.

66 Chai H, Farr RS, Froelich LA: Standardisation of bronchial inhalation challenge procedures. J Allergy Clin Immunol 1975;56:323–327.

67 Boushey HA, Holtzman M, Sheller JR, Nadel JA: Bronchial hyperreactivity. Am Rev Respir Dis 1980;121:389–413.

68 Cockcroft DW, Berscheid BA, Murdoch KY: Unimodal distribution of bronchial responsiveness to inhaled histamine in a random human population. Chest 1983;83:751–754.

69 Pattemore PK, Asher MI, Harrison AC, Mitchell EA, Rea HH, Stewart AW: The interrelationship among bronchial hyperresponsiveness, the diagnosis of asthma and asthma symptoms. Am Rev Respir Dis 1990;142:549–554.

70 Salome CM, Peat JK, Britton WJ, Woodcock AJ: Bronchial hyperresponsiveness in two populations of Australian schoolchildren. 1. Relation to respiratory symptoms and diagnosed asthma. Clin Allergy 1987;17:271–281.

71 Weiss ST, Tager IB, Weiss WJ, Munz A, Speizer FE, Ingram RH: Airway responsiveness in a population sample of adults and children. Am Rev Respir Dis 1984;129:898–902.

72 Hopp RJ, Bewtra AK, Nair NM, Townley RG: Specificity and sensitivity of methacholine inhalation challenge in normal and asthmatic children. J Allergy clin Immunol 1984;74:154–158.

73 Ten Velde GPM, Kreukniet J: The histamine inhalation provocation test and its reproducibility. Respiration 1984;45:131–138.

74 Josephs LK, Gregg I, Mullee MA, Holgate ST: Nonspecific bronchial reactivity and its relationship to the clinical expression of asthma: A longitudinal study. Am Rev Respir Dis 1989;140:350–357.

75 Sears MR, Holdaway MD, Silva P: Relationship between airway hyperresponsiveness, atopy and chilhood asthma: A longitudinal study. Am Rev Respir Dis 1987;135(suppl):A380.

76 Boulet LP, Cartier A, Hargreave FE: Asthma and increases in non-allergic bronchial responsiveness from seasonal pollen exposure. J Allergy Clin Immunol 1983;71:399–406.

77 Davies TRJ, Blainey AD: Occupational asthma; in Clark TJH, Godfrey S (eds): Asthma. London, Chapman & Hall, 1983, pp 202–241.

78 O'Connor GT, Weiss ST, Tager IB, Speizer FE: The effect of passive smoking on pulmonary function and nonspecific bronchial responsiveness in a population-based sample of children and young adults. Am Rev Respir Dis 1987;135:800–804.

79 Stempal DA, Boucher RC: Respiratory infection and airway reactivity. Med Clin North Am 1981; 65:1045–1053.

80 Clifford R, Pugsley A, Radfor M, Holgate ST: Symptoms, atopy and bronchial response to methacholine in parents with asthma and their children. Arch Dis Child 1987;62:66–73.

81 Ford LD, Bewtra AK, Coleman R, Herbers L, Townley RG: Methacholine challenge as a predictor for asthma. J Allergy Clin Immunol 1984;73:125.

82 Hopp RJ, Bewtra AK, Biven NM, Nair M, Townley RG: Bronchial reactivity pattern in nonasthma parents of asthmatics. Ann Allergy 1988;61:184–186.

83 Longo GR, Strinati R, Poli F, Fumi F: Genetic factors in bronchial hyperreactivity. Am J Dis Child 1987;141:331–334.

84 Sibbald B, Horn MEC, Brian EA, Gregg I: Genetic factors in childhood asthma. Thorax 1980;35: 671–674.

85 Van Herwerden L, Harrap S, Wong Z, Abramson MJ, Kutin JJ, Forbes AB, Raven J, Lanigan A, Haydn-Walters E: Linkage of the high-affinity IgE receptor gene with BHR even in the absence of atopy. Lancet 1995;346:1262–1265.

86 Konig E, Godfrey S: Exercise-induced bronchial lability in monozygotic (identical) and dizygotic (non-identical) twins. J Allergy Clin Immunol 1974;54:280–287.

87 Howard P: Respiratory Medicine in Clinical Practice. London, Arnold, 1991.

88 Gross NJ: What is this thing called love – or defining asthma. Am Rev Respir Dis 1980;121: 203–204.

89 Agosti JM, Sprenger JD, Lum LG: Transfer of allergen-specific IgE-mediated hypersensitivity with allogenic bone marrow transplant. N Engl J Med 1988;319:1623–1628.

90 Corris PA, Dark JH: Aetiology of asthma: Lessons from lung transplantation. Lancet 1993;341: 1369–1371.

91 Drinkwater H: Mendelian heredity in asthma. Br Med J 1909;i:88.

92 Meltzer SJ: Bronchial asthma as a phenomenon of anaphylaxis. JAMA 1910;55:1021–1024.

93 Meyers D, Bleecker ER: Approaches to mapping genes for allergy and asthma. Am J Respir Crit Care Med 1995;152:411–413.

94 Leigh D, Marley E: Bronchial asthma, a genetic population and psychiatric study. Oxford, Pergamon Press, 1967.

95 Lewitter FL, Tager LB, McGue M, Tishler PV, Speizer FE: Genetic and environmental determinants of level of pulmonary function. Am J Epidemiol 1984;120:518–529.

96 Horwood LJ, Ferguson DF, Shannon FT: Social and familial factors in the development of early childhood asthma. Paediatrics 1985;75:859–868.

97 Magnusson CGM: Cord serum IgE in relation to family history under the predictor of atopic disease in early infancy. Allergy 1988;43:241–251.

98 Carter CO: Genetics of common disorders. Br Med Bull 1969;25:52–57.

99 Lubs MLE: Empiric risks for genetic counselling in families with allergy. J Paediatr 1972;80:26–31.

100 Duffy DL, Martin NG, Battistutta D, Hopper JL, Mathews JD: Genetics of asthma and hay fever in Australian twins. Am Rev Respir Dis 1990;142:1351–1358.

101 Nieminen MM, Kaprio J, Koskenvou M: A population-based study of bronchial asthma in adult twin pairs. Chest 1991;100:70–75.

102 Morrison-Smith J: The prevalence of asthma and wheezing in children. Br J Dis Chest 1976;70:73.

103 Morrison-Smith J, Cooper S: Asthma and atopic disease in immigrants from Asia and the West Indies. Postgrad Med J 1981;57:774–776.

104 Ng TP, Hui KP, Tan WC: Prevalence of asthma and risk factors among Chinese, Malay, and Indian adults in Singapore. Thorax 1994;49:347–351.

105 Newman-Taylor A: Environmental determinants of asthma. Lancet 1995;345:296–299.

106 Waite DA, Eyles EF, Tonkin SL, O'Donnell TV: Asthma prevalence in Tokelauan children in two environments. Clin Allergy 1980;10:71–75.

107 Van Niererk CH, Weinberg EG, Shaw EG, Heese HDEV, Schalioyk DJ: Prevalence of asthma: A community study of urban and rural Xhosa. Clin Allergy 1979;9:319–324.

108 Brown P, Carlton D: Acute and chronic pulmonary airway disease in Pacific Island Micronesians. Am J Epidemiol 1978;108:266–273.

109 Burroughs B, Hallonen M, Barbee RA, Lebovitz MD: The relationship of serum immunoglobulin E to cigarette smoking. Am Rev Respir Dis 1981;124:523–525.

110 Anonymous: Smoking: Occupational allergic lung disease. Lancet 1985;i:965.

111 Venables KN, Topping MD, Howe W, Newman-Taylor AJ: Interaction of smoking and atopy in producing specific IgE antibody against hapten protein conjugate. Br Med J 1985;290:201–204.

112 Venables KM, Daly MB, Nunn AJ, Stevens JF, Farrer N, Hunter JV, Stewart M, Hughes EG, Newman-Taylor AJ: Smoking and occupational allergy in a platinum refinery. BMJ 1989;299: 939–942.

113 Magnusson CGN: Maternal smoking influences serum IgE and IgD levels and increases the risk for subsequent infant allergy. J Allergy Clin Immunol 1986;78:898–904.

114 Weiss ST, Tager IB, Munzo A, Speizer FE: The relationships of respiratory infections in early childhood to the occurrence of increased levels of bronchial responsiveness and atopy. Am Rev Respir Dis 1985;131:573–578.

115 Kjellman N: Effect of parental smoking on IgE levels in children. Lancet 1981;i:993–994.

116 Schachter EN, Witek TJ, Berk GJ, Hosein HR, Colic G, Caen W: Airway effects of low concentration of sulphur dioxide dose-response characteristics. Arch Environ Health 1984;39:34–42.

117 Molfino NA, Wright SC, Katz I, Tarlo S, Silverman F, McClean PA, Szalai JP, Raizenne M, Scutsky AS, Zamel N: Effect of low concentrations of ozone on inhaled allergen responses in asthmatic subjects. Lancet 1991;338:199–203.

118 Anto JM, Sunya J, Rodrigez-Royson R, Vazquez L: Community outbreak of asthma associated with inhalation of soya bean. N Engl J Med 1989;320:1097–1102.

119 Audman D: An outbreak of bronchial asthma in South Africa affecting more than two hundred persons caused by castor bean dust from an oil-pressing factory. Int Arch Allergy 1955;7:10–24.

120 Woolcock AH, Green W, Alpers MP: Asthma in a rural highland area of Papua New Guinea. Am Rev Respir Dis 1981;123:565–567.

121 Vanto T, Koivikko A: Dog hypersensitivity in asthmatic children. Acta Paediatr Scand 1983;72: 571–575.

122 Salter H: Diseases of the chest. Lancet 1870;i:183–187.

123 Kramer MS: Does breast feeding help protect against atopic disease? Biology, metholdology and a golden jubilee of controversy! J Paediatr 1988;112:181–190.

124 Burney P: A diet rich in sodium may potentiate asthma. Chest 1987;91:1435–1438.
125 Burney P, Britton JR, Chin NS, Tattersfield AE, Platt HS, Papacosta AO, Kelson MC: Response to inhaled histamine and twenty-four hour sodium excretion. Br Med J 1986;292:1483–1486.
126 Burney P, Nield JE, Twort CHC: The effect of changing dietary sodium on the bronchial response to histamine. Thorax 1989;44:36–41.
127 Britton JR, Pavord I, Richards K, Knox A, Wisnieski A, Weiss S, Tattersfield A: Dietary sodium intake and the risk of airway hyperreactivity in a random adult population. Thorax 1994;49:875–880.
128 Britton J, Pavord I, Richards K, Wisnieski A, Knox A, Lewis S, Tattersfield A, Weiss S: Dietary magnesium, lung function, wheezing and airway hyperreactivity in a random population sample. Lancet 1994;344:357–362.
129 Bloch H, Silverman R, Mancherje N, Grant S, Jagminas L, Scharf SM: Intravenous magnesium as an adjunct in the treatment of acute asthma. Chest 1995;107:1576–1581.
130 Tiffany BR, Berk WA, Todd IK, White SR: Magnesium bolus or infusion fails to improve expiratory flow in acute asthma exacerbation. Chest 1993;104:831–834.
131 Gordon RR: Breast feeding in eczema/asthma. Lancet 1982;i:910.
132 Hide DW, Guyer BM: Clinical manifestations of allergy related to breast and cow's milk feeding. Arch Dis Child 1981;56:172–175.
133 Saarinen U, Kajosaari M: Breast feeding as prophylaxis against atopic disease: Prospective follow-up study. Lancet 1995;346:1065–1069.
134 Godfrey KM, Barker DJP, Osman C: Disproportionate foetal growth and raised IgE concentrations in adult life. Clin Exp Allergy 1994;24:641–648.
135 Chandra RK: Interactions between early nutrition and the immune system; in Bock JL (ed): The childhood environment and adult disease. Ciba Found Symp No 156. Chichester, Wiley, 1991, pp 77–92.
136 Warner JA, Miles EA, Jones AC, Quint DJ, Colwell BM, Warner JO, et al: Is deficiency of interferon-gamma production by allergen-triggered cord blood cells a predictor of atopic excema? Clin Exp Allergy 1994;24:423–430.
137 Tang MLK, Kemp AS, Thorburn J, Hill DJ: Reduced interferon-γ secretion in neonates and subsequent atopy. Lancet 1994;344:983–985.
138 Pearson DJ, Freid DLJ, Taylor B: Respiratory allergy and month of birth. Clin Allergy 1980;10:585–591.
139 Bjork ST, Suoniemi I, Kiski V: Neonatal birch pollen contact and subsequent allergy to birch pollen. Clin Allergy 1980;10:585–591.
140 Bjorkston F, Suoniemi I: Time and intensity of first pollen contacts and risks of subsequent pollen allergy. Acta Med Scand 1981;209:299–303.
141 Frick OL, German DF, Mills J: The development of allergy in children. 1. Association with virus infections. Allergy Clin Immunol 1979;63:228–241.
142 Sven DK, Arora S, Gupta S, Sanyall RK: Studies of adrenergic mechanisms in relation to histamine sensitivity in children immunised with Bordatella pertussis vaccine. J Allergy Clin Immunol 1974; 54:25–31.
143 Agabiti N, Forastiere F, Porta D, Dell'Orco V, Corbo GM, Pistelli R: Socioeconomic status, number of siblings, and respiratory infections in early life as determinants of atopy in children. Eur Resp J 1995:8(suppl. 19), pp 197s, 1s–609s.
144 Devereux G, Ayatollahi SMT, Bromly CL, Stenton SC, Hendrick DG: Atopy is associated with family size and birth order. Eur Resp J 1995:8(suppl. 19), pp 197s, 1s–609s.

J.C. Dewar, Department of Therapeutics, University Hospital, Queen's Medical Centre, Nottingham NG7 2UH (UK)

Hall, IP (ed): Genetics of Asthma and Atopy.
Monogr Allergy. Basel, Karger, 1996, vol 33, pp 35–52

....

Animal Models of Bronchial Hyperreactivity

Charles Emala, Carol Hirshman

The Johns Hopkins University, Baltimore, Md., USA

Introduction

Asthma most likely represents a complex set of disorders grouped together into one category based on the symptom of episodic wheezing or reversible airway obstruction. Any attempt to dissect this disease or perhaps this group of diseases must include simplified approaches to investigate individual features of this disorder. Bronchial hyperreactivity, the exaggerated bronchoconstrictor response to a variety of stimuli, is an essential feature of human asthma. Thus, a substantial number of experimental animal models of airway hyperreactivity have been developed over the last few decades. These models largely fall into two broad categories: those in which the bronchial hyperreactivity is induced transiently in otherwise nonhyperreactive animals by specific pharmacologic or immunologic means, and those in which chronic bronchial hyperreactivity is present as a familial, presumably genetically-determined trait, simulating the situation of the human disease. While both approaches can be used to study the effects of interventions, the latter approach also promises to reveal possible genetic mechanisms underlying this aspect of asthma and is the focus of this chapter.

Bronchial hyperreactivity to agonists such as methacholine or histamine varies substantially between humans with active asthma and nonasthmatic individuals. A wide range of bronchial reactivity to these same agonists has been seen in many other species. In order to study the relative contribution of genetic factors to bronchial hyperreactivity, investigators have screened homogeneous strains of inbred animals for bronchial hyper- and hyporeactivity or have selectively bred and subsequently inbred animals showing exceptional bronchial reactivity with the eventual intent of identifying the genes controlling this trait.

The inbred laboratory mouse, because of the low level of variability within strains, the low cost, and the ease of breeding, is a well-described genetic resource. In contrast, rats of available strains do not necessarily carry the same allele at a particular locus on homologous chromosomes and less is known about their genetic make-up. However, their larger size allows more sophisticated physiological measurements. Inbred dogs, such as the basenji-greyhound (BG) dog model of airway hyperreactivity, are not homozygous at all loci, but the model is very well characterized with respect to the physiology of the airway and biochemistry of the airway smooth muscle. The large size of the animal allows access to homogenous airway tissue but little is known about the canine genome.

The Basenji-Greyhound Dog Model of Bronchial Hyperreactivity

The BG dog may be particularly useful in unraveling the complicated relationship between asthma, allergy, and bronchial hyperreactivity. The dog, like the human, spontaneously develops a disease syndrome related to aeroallergens such as ragweed pollen [1]. However, rhinitis, conjunctivitis, and especially dermatitis [2], rather than respiratory symptoms, predominate. Airway constriction can be induced experimentally in dogs natively allergic to antigen [3] as a result of a prior parasitic infection [4] or injection of antigen over a 16-week period just after birth [4]. Most investigators have been unsuccessful at sensitizing unselected adult mongrel dogs to specific antigens.

Antigen-induced airway constriction in the dog is mediated by an IgE-like antibody [5, 6]. Both airway [6] and skin responses [7, 8] can be transferred by passive immunization from allergic to nonallergic dogs. Moreover, IgE-like specific antibodies are found in respiratory secretions following antigen exposure [9]. Antigen aerosol challenge in sensitized dogs is associated with an increase in lung resistance and a decrease in lung compliance [3]. Lung resistance increases within 3–5 min, peaks at 10–15 min, and returns to baseline within a few hours. A delayed bronchoconstrictor response may recur a number of hours later [10, 11], and is associated with increases in cholinergic reactivity [12].

The dominant neural pathway controlling airway smooth muscle tone and airway caliber in the canine lung is the parasympathetic nervous system, the fibers of which run in the vagus nerve. The predominant bronchodilator system is the adrenergic system, via actions of circulating catecholamines [13, 14]. The dog does not have nonadrenergic noncholinergic bronchodilator nerves.

Pharmacologic agonists such as cholinergic agents, histamine and sero-tonin, bronchoconstrict the canine airway. Histamine and serotonin trigger bronchoconstriction by stimulation of vagal reflex pathways [15, 16], as well as by directly activating histamine H_1 and serotonin 5-HT$_2$ receptors on airway smooth muscle [17] and activating the inositol phosphate pathway. In contrast, cholinergic agonists such as carbachol directly activate muscarinic receptors on the airway smooth muscle with no reflex component. In canine trachealis muscle, M_2 and M_3 muscarinic receptors are found in a 9:1 ratio [18, 19]. The M_3 muscarinic receptor couples to the activation of the inositol phosphate pathway and muscle contraction while the M_2 muscarinic receptor couples to the inhibition of adenylyl cyclase [20, 21]. Thus, stimulation of the M_2 muscarinic pathway inhibits relaxation [18] induced by agonists such as isoproterenol or PGE$_1$, which are coupled to activation of adenylyl cyclase.

The dose of histamine or methacholine needed to produce a defined pulmonary response varies widely in individual dogs. In a large population of mongrel dogs, sensitivity to aerosol histamine was found to be log normally distributed with a 40-fold range [22]. Similar results were found with carbachol [23]. Bronchial reactivity to cholinergic agonists and histamine is closely corre-lated in the dog [23], as in the human [24], despite pharmacologic dissimilarities between histamine and cholinergic agonists and their predilections for different sites of action in the airway. This indicates that bronchial hyperrectivity in the dog is not restricted to responses initiating airway constriction by a specific receptor.

Because asthma is a disease that clusters in families [25, 26], and because antigen aerosol challenge had been the basis of most previous animal models, we originally selected from a colony of basenji dogs, crossed with greyhounds, three offspring that were unusually sensitive to *Ascaris summ* antigen aerosols [27]. These offspring were female siblings and were bred back to their father, and this eventually produced a colony of allergic dogs with chronic nonspecific bronchial hyperreactivity. In contrast to the mouse and the rat models of bronchial hyperractivity, which are agonist-specific, BG dogs demonstrate bronchial hyperreactivity to aerosols of methacholine [28, 29], histamine [30], calcium chelators [28], leukotriene D$_4$ [30] and hypotonic aerosols [31]. This hyperreactivity is not seen in purebred basenjis, purebred greyhounds or unse-lected mongrel dogs. As in patients with asthma, large volume ventilation results in bronchoconstriction in BG dogs, but results in bronchodilation in mongrels [32]. The β-adrenergic antagonist, propranolol increases bronchial reactivity to bronchoconstrictor agents in BG dogs, whereas it is without effect in mongrels [33, 34]. Moreover, β-adrenergic agonists are significantly less effective in protecting against the development of cholinergic constriction in the BG compared to the mongrel dog [35]. This difference was also observed in the small airways of the lung periphery [36].

Allergy in the BG dog is manifested by multiple positive skin tests to allergens common to their native environment [8], by chronic relapsing dermatitis [37], and by the presence of circulating hypodense eosinophils [38]. The allergic component involves IgE-like antibodies [5] and can be passively transferred to nonallergic dogs [8]. The bronchoalveolar lavage fluid of BG dogs contains greater numbers of mast cells [39] and eosinophils [40] than nonallergic control dogs. Moreover, the mast cells contained in the bronchoalveolar lavage fluid show higher spontaneous histamine release and histamine releasability [41] than mast cells from control animals.

To understand the role of familial factors in nonspecific bronchial hyperreactivity, breeding studies were conducted in BG dogs and in a control population of purebred basenji which were also allergic but which lacked bronchial hyperreactivity. The parent dogs were age-matched and were bred at the same time. Seven BG offspring and 4 basenji offspring were born within 10 days of each other and were raised in adjoining cages in the same room for 1 year, after which time they were tested with aerosols of methacholine and of allergens common in their environment. All BG and basenji offspring had at least two immediate wheal skin test responses equal to or greater than the positive control, indicating that offspring from both breedings were, like their parents, allergic. In contrast, BG offspring, like their parents, showed more than 10-fold greater sensitivity to methacholine than basenji offspring (fig. 1) [42]. This study suggests that in the BG dog nonspecific bronchial hyperreactivity may, in part, be genetically determined separate from allergy with a dominant mode of inheritance.

Increases in airway smooth muscle tone which decrease the caliber of the airway and contribute to increases in bronchial reactivity may be due to increased activity of the signaling pathways regulating contraction or decreased activity of pathways mediating relaxation. The major pathway regulating relaxation of the airway smooth muscle of the dog involves circulating epinephrine activating β_2-adrenergic receptors, which couple to $G_{s\alpha}$, stimulate adenylyl cyclase, and increase intracellular cAMP.

To test the hypothesis that an abnormal biochemical pathway in the airway smooth muscle plays a role in the bronchial hyperreactivity of the BG dog, 5 BG offspring (allergic and hyperreactive) and 5 purebred basenjis (allergic but not hyperreactive) at 2 years of age, were matched with a control group of 5 2-year-old purebred greyhounds. The greyhounds had no positive skin wheal responses to antigens common in their environment, and were significantly less reactive than BG dogs to aerosols of either histamine or methacholine (fig. 2) [29]. The dogs were sacrificed and tracheal and bronchial muscle preparations

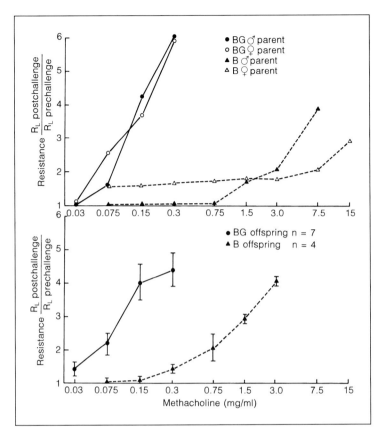

Fig. 1. Increase in pulmonary resistance (R_L) to incremental doses of methacholine in basenji-greyhound (BG) and basenji (B) parents (top) and BG and B offspring (bottom). In offspring each point represents mean $\pm$ SE of 7 BG or 4 B dogs. [Reprinted from 42, with permission.]

were evaluated with respect to contractile and relaxant responses. No animal had previously received β-adrenergic agonist or corticosteroid treatment, nor had any received aerosol challenges in the two months prior to sacrifice. Airway smooth muscle obtained from the BG dog was less sensitive to isoproterenol, when the muscle was precontracted with methacholine, than was similar muscle from the greyhounds [43] or from the basenjis [44]. In additional studies, basenjis, which were similarly allergic but lacked bronchial hyperreactivity, had airway smooth muscle relaxation responses which resembled those of the nonallergic greyhounds or mongrels with respect to β-adrenergic agonists [45]. These

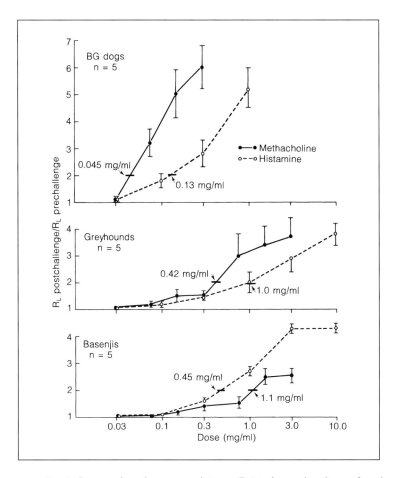

Fig. 2. Increase in pulmonary resistance (R_L) to increasing doses of methacholine and histamine. The dose of drug eliciting a 200% increase in R_L in BG, basenji, and greyhound dogs is indicated in each graph. [Reprinted from 29, with permission.]

data suggest that the decreased sensitivity to β-adrenergic agonists was related in some way to the bronchial reactivity of the BG dog.

The defect in the airway smooth muscle of the BG dog is β-adrenergic receptor-specific, since decreased generation of cAMP was not seen with agents which bypassed th β-adrenergic receptor [45]. This selective β-adrenergic receptor impairment occurred despite increased numbers of β-adrenergic receptors [45]. Agonist and antagonist binding to the β-adrenergic receptor in the lung membranes of the BG dog were not impaired but more of the β-adrenergic receptors in the BG lung were in an inactive conformation [46]. A qualitatively similar β-adrenergic impairment was seen in leukocytes [47] and tracheal epithelial cells from this model [48].

Does this β-adrenergic receptor deficit in the BG dog represent a mutation in the gene encoding the $β_2$-adrenergic receptor, since most of the β-adrenergic receptors in the lung and in the leukocytes are of the $β_2$ subtype? Two lines of evidence suggest that the β-adrenergic receptor dysfunction in the BG dog tissues is acquired. First, cultured airway smooth muscle cells from the BG dog are indistinguishable from those of mongrel dogs with respect to β-adrenergic receptor-activated adenylyl cyclase generation [49]. Thus, the defect does not persist in culture, suggesting that some factor or factors in the lung surrounding the airway smooth muscle cells of the BG dog are inducing the β-adrenergic receptor dysfunction. Second, the DNA sequence of the $β_2$-adrenergic receptor gene coding for regions of the receptor protein involved in coupling to $G_{sα}$ (the third intracellular loop and carboxy-terminal tail) is identical in the BG and mongrel dog.

Bronchial hyperreactivity is not associated with increased in vitro trachealis muscle sensitivity to histamine [43], serotonin [44], or methacholine [43, 45] in the BG dog. This is in agreement with in vitro studies of airway smooth muscle from humans with asthma, which showed no correlation with in vivo responses to contractile agonists [50–53]. A recent study from our laboratory demonstrating a 3- to 4-fold increase in the number of muscarinic receptors in BG trachealis muscle appeared initially to be inconsistent with these results [54]. However, immunoprecipitation with muscarinic subtype-specific antibodies revealed that the entire increase in total muscarinic receptor number was attributable to an increase in the M_2 muscarinic receptor subtype. Muscarinic M_3 receptor numbers were not different in BG and control muscle, which is consistent with the in vitro studies showing that methacholine sensitivity was not increased [43–45]. Furthermore, trachealis muscle from the BG dog showed increased cholinergic inhibition of forskolin-stimulated adenylyl cyclase [54]. Thus, abnormalities in signal transduction pathways of airway smooth muscle leading to both impaired generation of cAMP and increased inhibition of cAMP formation are important functionally in this model. Whether the M_2 muscarinic receptor up-regulation is the result of cross-regulation by the $β_2$-adrenergic pathway or a genetic defect in the regulation of M_2 muscarinic receptor expression remains to be determined.

Rodent Models of Bronchial Hyperreactivity

If the inheritance of bronchial reactivity is polygenic, bronchial hyperreactivity is probably determined by multiple genes interacting with environmental factors. Thus, our ability to understand the contribution of individual genes to the phenotype of bronchial reactivity is limited in outbred populations in which both hereditary and environmental factors are uncontrolled. Rodents such as rats and mice are particularly suited for genetic studies because of the short gestation period, the lower cost of housing, the ability to control environmental factors, and the ability to produce inbred strains of animals [55].

The major objective of inbreeding is to reduce genetic variability to near zero by eliminating heterozygosity at every genetic locus. An inbred strain of rodent is produced by 20 or more consecutive generations of brother-sister mating [56]. This results in animals that are essentially homozygous at all genetic loci. Inbreeding provides investigators with a theoretically unlimited supply of tissue from genetically identical animals in which environmental factors can be more easily controlled. Thus, biological variability for any physiologic response within a strain can be attributed to environmental influences while variability between strains can be attributed to genetic differences.

The Rat

Although the rat is difficult to sensitize to foreign proteins [57, 58], systemic anaphylaxis can be produced in the rat by appropriate challenge in animals sensitized to foreign proteins using adjuvants such as *Bordetella pertussis*. The rat, like the human, produces IgE in response to active immunization, and the amount of IgE produced is strain-specific [59]. The Brown Norway (BN) rat strain produces high levels of IgE in response to ovalbumin sensitization [60, 61] and develops both early and late responses after inhalational challenge [62]. The pulmonary response to antigen in the rat differs from that in the human and in most nonrodent animal models of asthma in that it is mediated primarily by serotonin and is effectively prevented with serotonin receptor antagonists [63–65].

The rat differs dramatically from most other laboratory animals in that histamine, even in massive doses, does not contract airway smooth muscle either in the intact animal [64] or in isolated preparations of trachea, bronchus, or lung parenchyma [66, 67]. Histamine, released from the lung and other sites during anaphylaxis [68], contributes to systemic responses with little or no stimulatory effect on airways. Conversely, serotonin, which has little effect in human airways, produces bronchoconstriction in the rat. Serotonin bronchoconstricts the rat airway [64, 69] by direct effects on airway smooth muscle [66], since neither bilateral vagotomy [70] nor atropine pretreatment [71] influence the bronchoconstrictor response. The rat differs in this respect from both the dog, in which part of serotonin-induced bronchoconstriction is vagally-mediated [16], and the mouse, in which most of the response is vagally-mediated [72, 73].

Cholinergic agonists (methacholine, carbachol, and acetylcholine), like serotonin, have little or no irritant effects but directly activate cholinergic receptors on the airway smooth muscle. In the rat, radioligand binding and immunoprecipitation studies identified M_2 and M_3 muscarinic receptors in central airways while M_1, M_2 and M_3 muscarinic receptors were found in peripheral lung [74, 75]. The M_2 subtype predominates throughout the airways [74].

Genetic studies of bronchial hyperreactivity in the rat have identified strain-related differences in bronchial reactivity to specific agonists, and have evaluated patterns of inheritance by phenotypically characterizing progeny of hyperreactive and hyporeactive strains. Eidelman et al. [76] found that Fisher 344 rats were significantly more reactive to methacholine aerosols than Lewis (LE) rats. Pauwels et al. [71] compared bronchial reactivity to both serotonin and carbachol and found that IC rats were hyperreactive to both serotonin and carbachol. RA rats were intermediate to both while OM/N rats were hyperreactive to serotonin but hyporeactive to carbachol. BN and LE rats were hyporeactive to both. Moreover, reactivity to carbachol and serotonin was not related. When a strain with bronchial hyperreactivity to serotonin (IC) was bred with a strain which was hyporeactive to serotonin (DA) and airway responses were measured in the progeny, all the rats of the first filial (F_1) generation exhibited low bronchial reactivity to serotonin, but backcrosses between the F_1 generation and the IC rats resulted in rats with large variability in their serotonin reactivity which was interpreted as an autosomal recessive pattern of inheritance [71]. Thus, in contrast to the human asthmatic and the BG dog in whom bronchial hyperreactivity is nonspecific, bronchial hyperreactivity in inbred rats is agonist-specific.

Reactivity to serotonin was further investigated following endotoxin aerosol challenge in three rat strains. BN, Fisher 344, and RA strains were exposed to endotoxin and bronchial reactivity to intravenous serotonin was studied 90 min later. Although all three strains developed an inflammatory response characterized by increased numbers of neutrophils in the bronchoalveolar lavage fluid, only the RA and Fisher 344 strains developed bronchial hyperreactivity to serotonin [77]. This suggests that the genetic determinants of bronchial hyperreactivity to serotonin in the rat strains studied are independent of factors regulating neutrophil influx, and that the presence of inflammation alone does not determine serotonin bronchial reactivity in the rat.

Piechuta et al. [63] identified three strains of rats that had markedly different airway responses to aerosolized ovalbumin after sensitization. Selective inbreeding of one strain (Sprague-Dawley) resulted in an increase in the incidence of dyspnea induced by ovalbumin aerosols from 44% in the parental strains to 55% in the F_1 generation and to greater than 90% in the second filial (F_2) generation ($F_1 \times F_1$) and third filial (F_3) generation ($F_2 \times F_2$), suggesting that antigen-induced dyspnea in the rat is controlled genetically [78]. The selectively inbred and nonsensitized Sprague-Dawley rats were also more reactive to serotonin [79] and citric acid aerosols [80] than control Sprague-Dawley rats. Airway smooth muscle from the inbred strain was more sensitive to serotonin and leukotriene D_4 than similarly sensitized control rats [79]. However, methacholine sensitivity was similar in inbred and control Sprague-Dawley rats.

Thus, inbreeding of rats encourages genetic selection of a gene or genes that confer airway hyperreactivity to serotonin and leukotriene LTD_4. However, reactivity to methacholine must involve different genetic loci in this model. Specific linkage studies have not yet been performed. It is also unclear whether the rat strains that are referred to as 'inbred' have actually developed from at least 20 generations of brother-sister matings to obtain homozygosity at all genetic loci.

The Inbred Mouse

The mouse is similar to the rat in its airway responses to specific contractile agonists. The mouse airway is unreactive to histamine in vivo [81] and in vitro [82], while serotonin is a potent bronchoconstrictor [81, 83]. Although serotonin is believed to have both direct and indirect effects on the airways of most species, in the mouse serotonin-induced bronchoconstriction is totally inhibited by atropine [81, 83]. This suggests that most of the airway response to serotonin is mediated by neural pathways in the mouse. In view of this finding, it is somewhat surprising that serotonin-induced bronchoconstriction was totally inhibited by a $5\text{-}HT_2$ receptor antagonist in a recent study [73], since $5\text{-}HT_3$ serotonin receptors are the subtype found on peripheral nerves. Cholinergic agonists (carbachol, methacholine, acetylcholine) [81, 83, 84] act directly on muscarinic receptors of the airway smooth muscle to constrict the airways, presumably by activating M_3 muscarinic receptors on airway smooth muscle. Norepinephrine also bronchoconstricts the mouse airway [81], while prostaglandins D_2 and $F_{2\alpha}$ [81], leukotrienes C_4 and D_4 [81] and platelet-activating factor [84] are without direct effects.

In human bronchial smooth muscle, the β_2 subtype is the predominant β-adrenergic receptor [85, 86] mediating bronchodilatation, while smooth muscle of the mouse trachea contains predominantly β_1-adrenergic receptors [87]. The potential importance of the β-adrenergic receptor in mouse airway responsiveness is supported by the finding that propranolol pretreatment enhances responsiveness to serotonin, suggesting that baseline β-adrenergic activity exists in the mouse airway [81].

Inbred strains of mice show wide variability in bronchial reactivity to challenge agonists (fig. 3). Methods to measure these responses include an

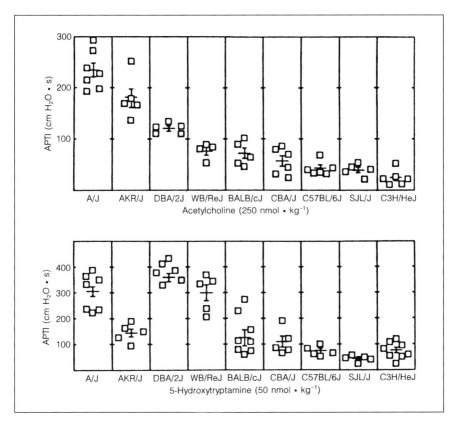

Fig. 3. The effects of acetylcholine or serotonin on the airway pressure time index (APTI) in nine inbred strains of mice. APTI is the integrated change in peak airway pressure over time in response to agonist and is variable between strains. □ = APTI for an individual mouse; ± = mean ± SE. [Reprinted from 72, with permission.]

integral measurement of the change in peak airway pressure over time (APTI) [83], the measurement of respiratory system resistance and elastance [88], and the measurement of pulmonary conductance and compliance [81]. Agreement exists between APTI and respiratory resistance measures [89]. Among the inbred mouse strains tested, the A/J mouse is hyperreactive to cholinergic agonists [83, 90] and to serotonin [72], while the C3H/HeJ and C57BL/6J [72] strains are hyporeactive to these agonists. The DBA/2J mouse strain, on the other hand, is hyperreactive to serotonin but not to acetylcholine [72], while the AKR/J strain [72, 84] is hyperreactive to acetylcholine but not to serotonin.

To identify genes that determine bronchial reactivity, genetic crosses between inbred strains that significantly differ in their bronchial reactivity to an agonist are carried out to produce the F_1 generation [55]. These F_1 animals are then mated to each other to produce the F_2 generation or back to each of the parental strains to produce the first backcross generations. The offspring from the F_2 and backcross generations thus differ at numerous genetic loci along each chromosome, which will segregate independently at meiosis. Conclusions are drawn about the genetic basis for the difference in airway reactivity between two strains on the basis of the frequency of mice in the segregating generations that resemble each parent strain or possess an intermediate response with respect to bronchial reactivity to an agonist or perhaps to a number of challenge agonists.

Using these classic methods of genetic analysis, Levitt and Mitzner [72, 83] bred A/J mice and C3H/HeJ mice to evaluate bronchial reactivity to acetylcholine and to serotonin in the F_1 and F_2 mice. They found that the F_1 mice were hyporeactive and that approximately 25% of the F_2 mice resembled the A/J progenitor strain and were hyperreactive to acetylcholine [83] or serotonin [72]. These studies are consistent with the hypothesis that bronchial hyperreactivity is inherited as a simple autosomal recessive trait. If the gene or genes that determine bronchial hyperreactivity to cholinergic agonists and serotonin are the same or are closely linked, these genes should co-segregate (be inherited together) in the F_2 mice. However, when a series of F_2 mice were evaluated for bronchial hyperreactivity to both acetylcholine and serotonin, the hyperreactive phenotype to acetylcholine did not co-segregate with the hyperreactive phenotype to serotonin [72] (fig. 4), suggesting that bronchial hyperreactivity to serotonin and to cholinergic agonists were determined by different genes and that bronchial reactivity in the mouse, like in the rat, is agonist-specific.

In contrast to these studies, De Sanctis et al. [90] found that the F_1 generation of a cross between hyperreative A/J mice and hyporeative C57 BL/6J mice resembled the A/J strain. Longphre and Kleeberger [84] had similar findings in a study crossing hyperreative AKR/J mice with hyporeactive C3H/HeJ, suggesting that bronchial hyperreactivity is inherited as a dominant rather than as a recessive trait. These two more recent studies and a third study by Ewart et al. [88], using A/J and C3H/HeJ strains, all demonstrated intermediate phenotypes in the F_2 generations. This suggests that bronchial reactivity to cholinergic agonists in inbred mice is more complex than simple dominant-recessive inheritance.

Further support for this view comes from two recent genetic studies in hyperreactive and hyporeactive mouse strains [88, 90]. Both studies used a genetic mapping or random marker typing approach. A large number of genetic markers or polymorphisms distributed across the genome that differentiate the two parental strains, but which are in themselves functionally

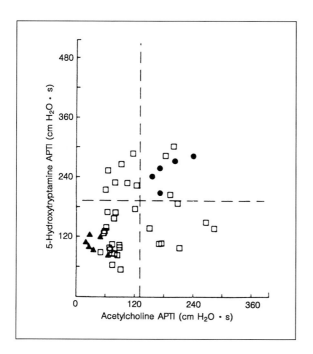

Fig. 4. Co-segregation analysis of bronchial hyperreactivity to serotonin and acetylcholine in hyperreactive mice (A/J), hyporeactive mice (C3H/HeJ) and their 36 F_2 offspring $(CAF_1) \times (CAF_1)$. Dashed lines represent apparent division between hyperreactive phenotypes based on parental (A/J and C3H/HeJ) responses. Airway responses to acetylcholine do not co-segregate with airway responses to serotonin. ● = A/J; ▲ = C3H/HeJ; □ = F_2 hybrids. [Reprinted from 72, with permission.]

meaningless, are identified. If one of the markers is found to co-segregate with bronchial hyperreactivity in an F_2 intercross or backcross, a specific chromosomal region of importance can be identified. Only markers that are physically close to the gene of interest will demonstrate a high degree of association or linkage. The likelihood that a given marker resides within a region of the genome close to the location of the gene of interest (in this case the bronchial hyperreactivity gene) is determined by calculating the logarithm of the odds ratio (LOD score). A LOD score of three (1:1,000) or greater is thought to indicate linkage.

If bronchial hyperreactivity is a continuous, rather than a discrete variable with multiple genes contributing to the phenotype in the mouse, the expression of the quantitative differences in phenotypes depends on genes whose effects may be modified by other genes or environmental factors. These are called quantitative trait loci (QTL). De Sanctis et al. [90] (using a backcross of hyperreactive A/J and hyporeactive C57 BL/6J) found significant linkage of the cholinergic bronchial hyperreactive phenotype to two loci – one on chromosome 2 and one on chromosome 15. In contrast, Ewart et al. [88] (using a backcross of hyperreactive A/J mice and hyporeactive C3H/HeJ mice) found significant linkage of the cholinergic bronchial hyperreactive phenotype to chromosome 6. The reasons for the different results found in the two studies may depend on the genetic make-up of the cross and the parameters chosen to phenotype the mice. These include the method of measuring cholinergic reactivity, the anesthetic used, the cholinergic agent selected, as well as the age, gender and previous environmental exposure of the animals. Although this reductionist strategy of selecting for strains for mice that are hyperreactive to only one agonist may simplify genetic linkage studies, by doing this, one may actually be biasing against the gene(s) responsible for the nonspecific nature of bronchial hyperreactivity found in human asthma.

While random marker typing represents a powerful technique to search for unknown disease genes, the identification of a genetic marker linked to cholinergic bronchial hyperreactivity represents only the first step in the identification of genes determining cholinergic bronchial hyperreactivity. At this point in time, the studies have identified rather large chromosomal segments as carrying one or more genes related to cholinergic bronchial hyperreactivity in the mouse. On the basis of these studies, it is premature to speculate on possible candidate genes for asthma, although these approaches may ultimately allow the identification and testing of candidate genes for asthma.

In summary: The familial incidence of asthma suggests a genetic basis for the disease. However, the genetic component of the asthma phenotype is difficult to study in human populations where the control of breeding and the environment are not possible. This has motivated the development of several animal models of bronchial hyperreactivity which allow specific genetic hypotheses to be tested. Each of the species chosen has both advantages and disadvantages for study of the genetic basis of bronchial hyperreactivity. While larger species allow easier and more reliable phenotypic characterization, smaller species allow better genotypic characterization.

The BG dog has many phenotypic similarities to human asthma including nonspecific bronchial hyperreactivity and allergy. Airway smooth muscle from the BG has a selective β-adrenergic impairment in vitro similar to that described in the human. Additionally, the β-adrenergic impairment in the BG, like in the human, appears to be an induced defect rather than a defect in the receptor protein per se. However, genetic study of the BG dog is hampered by the high cost of selective inbreeding, the longer gestation period, and the limited amount of information available concerning the dog genome.

The availability of inbred strains of mice has encouraged studies in this species in which a wealth of genetic information is currently available. Recent genetic studies in this relatively simple animal model system suggest that understanding the genetic determinants of bronchial hyperreactivity to even one agonist will not be easy. In contradistinction to monogenetic disorders, the identification of even a few genes relevant to bronchial hyperreactivity in humans will pose a major challenge. Progress in the understanding of the pathobiology of human asthma depends upon exploiting each of the animal models for their respective advantages. Insight gained from each model may allow the identification of specific candidate genes which can then be evaluated for their potential contribution to human asthma.

References

1   Patterson R: Investigation of spontaneous hypersensitivity of the dog. J Allergy 1960;31:351–363.
2   Anderson W: Atopic dermatitis in the dog. Cutis 1975;15:955–960.
3   Gold WM, Kessler GF, Yu DY, Frick OL: Pulmonary physiologic abnormalities in experimental asthma in dogs. J Appl Physiol 1972;33:496–501.
4   Kepron W, James JM, Kirk B, Sehon AH, Tse KS: A canine model for reaginic hypersensitivity and allergic bronchoconstriction. J Allergy Clin Immunol 1977;59:64–69.
5   Peters JE, Hirshman CA, Malley A: The basenji-greyhound dog model of asthma: Leukocyte histamine release, serum IgE, and airway response to inhaled antigen. J Immunol 1982;129:1245–1249.
6   Kesler GF, Frick OL, Gold WM: Immunologic and physiologic characterization of the role of reaginic antibodies in experimental asthma in dogs. Int Arch Allergy 1974;47:313–328.
7   Booth BH, Patterson R, Talbot CH: Immediate-type hypersensitivity in dogs: Cutaneous, anaphylactic, and respiratory responses to *Ascaris*. J Lab Clin Med 1970;76:181–189.
8   Hirshman CA, Leu DB, Austin DR, Chan SC, Veith L, Hanifin JM: Elevated mononuclear leukocyte phosphodiesterase in allergic dogs with and without airway hyperresponsiveness. J Allergy Clin Immunol 1987;79:46–53.
9   Patterson R, Roberts M, Pruzansky JJ: Comparisons of reaginic antibodies from three species. J Immunol 1969;102:466–475.
10   Sasaki H, Yanai M, Shimura S, Okayama H, Aikawa T, Sasaki T, Takishima T: Late asthmatic response to *Ascaris* antigen challenge in dogs treated with metyrapone. Am Rev Respir Dis 1987; 136:1459–1465.
11   Turner CR, Spannhake EW: Acute topical steroid administration blocks mast cell increase and the late asthmatic response of the canine peripheral airways. Am Rev Respir Dis 1990;141:421–427.
12   Itabashi S, Ohrui T, Sekizawa K, Aikawa T, Nakazzawa H, Sasaki H: Late asthmatic response causes peripheral airway hyperresponsiveness in dogs treated with metyrapone. Int Arch Allergy Immunol 1993;101:215–220.
13   Kneussl MP, Richardson JB: Alpha-adrenergic receptors in human and canine tracheal and bronchial smooth muscle. J Appl. Physiol 1978;45:307–311.
14   Leff AR, Munoz NM: Evidence for two subtypes of alpha-adrenergic receptors in canine airway smooth muscle. J Pharmacol Exp Ther 1981;217:530–535.
15   Vidruk EH, Hahn HL, Nadel JA, Sampson SR: Mechanism by which histamine stimulates rapidly adapting receptors in dog lungs. J Appl Physiol 1977;43:397–402.
16   Hahn HL, Wilson AG, Graf PD, Fischer SP, Nadel JA: Interaction between serotonin and efferent vagus nerves in dog lungs. J Appl Physiol 1978;44:144–149.
17   Chand N: Distribution and classification of airway histamine receptors: The physiological significance of histamine H_2-receptors. Adv Pharmacol Chemother 1980;17:103–131.

18   Fernandes LB, Fryer AD, Hirshman CA: M_2 muscarinic receptors inhibit isoproterenol-induced relaxation of canine airway smooth muscle. J Pharmacol Exp Ther 1992;262:119–126.

19   Emala CW, Aryana A, Levine MA, Yasuda RP, Satkus SA, Wolfe BB, Hirshman CA: Expression of muscarinic receptor subtypes and the relationship between m_2 receptors and inhibition of adenylyl cyclase in lung. Am J Physiol 1995;268:L101–L107.

20   Sankary RM, Jones CA, Madison JM, Brown JK: Muscarinic cholinergic inhibition of cyclic AMP accumulation in airway smooth muscle: Role of a pertussis-toxin sensitive protein. Am Rev Respir Dis 1988;138:145–150.

21   Jones CA, Madison JM, Tom-Moy M, Brown JK: Muscarinic cholinergic inhibition of adenylate cyclase in airway smooth muscle. Am J Physiol 1987;253:C97–C104.

22   Snapper JR, Drazen JM, Loring SH, Schneider W, Ingram RH: Distribution of pulmonary responsiveness to aerosol histamine in dogs. J Appl Physiol 1978;44:738–742.

23   Snapper JR, Drazen JM, Loring SH, Braasch PS, Ingram RH: Vagal effects on histamine, carbachol, and prostaglandin $F_{2\alpha}$ responsiveness in the dog. J Appl. Physiol 1979;47:13–16.

24   Spector SL, Farr RS: A comparison of methacholine and histamine inhalations in asthmatics. J Allergy Clin Immunol 1975;56:308–316.

25   Bray GW: The hereditary factor in asthma and other allergies. Br Med J 1930;i:384–387.

26   Townley RG, Bewtra A, Wilson AF, Hopp RJ, Elston RC, Watt GD: Segregation analysis of bronchial response to methacholine inhalation challenge in families with and without asthma. J Allergy Clin Immunol 1986;77:384–387.

27   Hirshman CA: Asthma. Comp Pathol Bull 1985;7:3–4.

28   Hirshman CA, Malley A, Downes H: The basenji-greyhound dog model of asthma: Reactivity to *Ascaris suum,* citric acid, and methacholine. J Appl. Physiol 1980;49:953–957.

29   Hirshman CA, Downes H: Airway responses to methacholine and histamine in basenji-greyhound and other purebred dogs. Respir Physiol 1986;63:339–346.

30   Hirshman CA, Darnell M, Brugman T, Peters J: Airway constrictor effects of leukotriene D_4 in dogs with hyperreactive airways. Prostaglandins 1983;25:481–490.

31   Osborne ML, Evans TW, Sommerhoff CP, Chung KF, Hirshman CA, Boushey HA, Nadel JA: Hypotonic and isotonic aerosols increase bronchial reactivity in basenji-greyhound dogs. Am Rev Respir Dis 1987;135:345–349.

32   Davidson AB, Hirshman CA, Downes H, Drazen JM: Large volume ventilation results in broncho-constriction in basenji-greyhound crossbred dogs. J Appl Physiol 1987;62:2308–2313.

33   Hirshman CA, Downes H, Leon DA, Peters JE: Basenji-greyhound dog model of asthma: Pulmonary responses after β-adrenergic blockade. J Appl Physiol 1981;51:1423–1427.

34   Tobias JD, Sauder RA, Hirshman CA: Pulmonary reactivity to methacholine during β-adrenergic blockade: Propranolol versus esmolol. Anesthesiology 1990;73:132–136.

35   Tobias JD, Sauder RA, Hirshman CA: Reduced sensitivity to β-adrenergic agonists in basenji-greyhound dogs. J Appl. Physiol 1990;69:1212–1219.

36   Lindeman KS, Hirshman CA, Freed AN: Functional antagonism of airway constriction in the canine lung periphery. J Appl Physiol 1991;71:1848–1855.

37   Butler JM, Peters JE, Hirshman CA, White CR, Margolin LB, Hanifin JM: Pruritic dermatitis in asthmatic basenji-greyhound dogs: A model for human atopic dermatitis. J Am Acad Dermatol 1983;8:33–38.

38   Brown RH, Hirshman CA: Hypodense eosinophils in peripheral blood of the basenji-greyhound dog model of airway hyperreactivity. Am Rev Respir Dis 1989;139:A480.

39   Hirshman CA, Austin DR, Klein W, Hanifin JM, Hulbert W: Increased metachromatic cells and lymphocytes in bronchoalveolar lavage fluid of dogs with airway hyperreactivity. Am Rev Respir Dis 1986;133:482–487.

40   Darowski J, Hannon VM, Hirshman CA: Corticosteroids decrease airway hyperresponsiveness in the basenji-greyhound dog model of asthma. J Appl Physiol 1989;66:1120–1126.

41   Hirshman CA, Austin DR, Kettelkamp NS: Enhanced bronchoalveolar lavage cell histamine releasability in allergic dogs with and without airway hyperresponsiveness. J Allergy Clin Immunol 1988; 81:829–835.

42   Hirshman CA, Downes H, Veith L: Airway responses in offspring of dogs with and without airway hyperreactivity. J Appl Physiol 1984;56:1272–1277.

43   Downes H, Austin DR, Parks CM, Hirshman CA: Comparison of drug responses in vivo and in vitro in airways of dogs with and without airway hyperresponsiveness. J Pharmacol Exp Ther 1986; 237:214–219.

44   Downes H, Austin DR, Parks CM, Hirsman CA: Comparison of in vitro drug responses in airways of atopic dogs with and without in vivo hyperresponsiveness. Pulm Pharmacol 1989;2:209–216.

45   Emala CW, Black C, Curry C, Levine MA, Hirshman CA: Impaired β-adrenergic receptor activation of adenylyl cyclase in airway smooth muscle in the basenji-greyhound dog model of airway hyperresponsiveness. Am J Respir Cell Mol Biol 1993;8:668–675.

46   Emala CW, Aryana A, Levine MA, Hirshman CA: Impaired relaxation of airway smooth muscle in asthma: Uncoupling of the β-adrenergic receptor. Anesthesiology 1994;81:A1444.

47   Emala CW, Levine MA, Aryana A, Margolick JB, Hirshman CA: Reduced adenylyl cyclase activation with no decrease in β-adrenergic receptors in basenji-greyhound leukocytes: Relevance to β-adrenergic responses in airway smooth muscle. J Allergy Clin Immunol 1995;95:860–867.

48   Croxton TL, Takahashi M, Hirshman CA: Decreased ion transport by tracheal epithelium of the basenji-greyhound dog. J Appl Physiol 1994;76:1489–1493.

49   Hungerford CL, Emala CW, Hirshman CA: Does impaired β-adrenergic stimulation of adenylyl cyclase in trachealis smooth muscle of the BG dog persist in cultured muscle cells? Am J Respir Crit Care Med 1994;149:A1083.

50   Bai TR: Abnormalities in airway smooth muscle in fatal asthma. Am Rev Respir Dis 1990;141: 552–557.

51   Cerrina J, Ladurie MLR, Labat C, Raffestin B, Bayol A, Brink C: Comparison of human bronchial muscle responses to histamine in vivo with histamine and isoproterenol agonists in vitro. Am Rev Respir Dis 1986;134:57–61.

52   Goldie RG, Spina D, Henry PJ, Lulich KM, Paterson JW: In vitro responsiveness of human asthmatic bronchus to carbachol, histamine, β-adrenoceptor agonists and theophylline. Br J Clin Pharmacol 1986;22:669–676.

53   Whicker SD, Armour CL, Black JL: Responsiveness of bronchial smooth muscle from asthmatic patients to relaxant and contractile agonists. Pulm Pharmacol 1989;2:25–31.

54   Emala CW, Aryana A, Levine MA, Yasuda RP, Satkus SA, Wolfe BB, Hirshman CA: Basenji-greyhound dog: Increased m_2 muscarinic receptor expression in trachealis muscle. Am J Physiol 1995;268:L935–L940.

55   Levitt RC, Mitzner W, Kleeberger SR: A genetic approach to the study of lung physiology: Understanding biological variability in airway responsiveness. Am J Physiol 1990;258:L157–L164.

56   Green EL: Biology of the Laboratory Mouse, ed 2. New York, McGraw-Hill, 1966.

57   Longcope WT: Insusceptibility to sensitization and anaphylactic shock. J Exp Med 1922;36:627–643.

58   Sanyal RK, West GB: Anaphylactic shock in the albino rat. J Physiol (Lond) 1958;142:571–584.

59   Murphey SM, Brown S, Miklos N, Fireman P: Reagin synthesis in inbred strains of rats. Immunology 1974;27:245–253.

60   Sapienza S, Eidelman DH, Renzi PM, Martin JG: Role of leukotriene D_4 in the early and late pulmonary responses of rats to allergen challenge. Am Rev Respir Dis 1990;14:353–358.

61   Renzi PM, Olivenstein R, Martin JG: Inflammatory cell populations in the large airways and small airways and parenchyma of rats after antigen challenge. Am Rev Respir Dis 1993;147:967–974.

62   Eidelman DH, Bellofiore S, Martin JG: Late airway responses to antigen challenge in sensitized inbred rats. Am Rev Respir Dis 1988;137:1033–1037.

63   Piechuta H, Smith ME, Share NN, Holme G: The respiratory response of sensitized rats to challenge with antigen aerosols. Immunology 1979;38:385–392.

64   Church MK: Response of rat lung to humoral mediators of anaphylaxis and its modification by drugs and sensitization. Br J Pharmacol 1975;55:423–430.

65   Stotland LM, Share NN: Pharmacological studies on active bronchial anaphylaxis in the rat. Can J Phsyiol Pharmacol 1974;52:1119–1125.

66   Burns JW, Doe JE: A comparison of drug-induced responses on rat tracheal, bronchial and lung strip in vitro preparations. Br J Pharmacol 1978;64:71–74.

67 Lulich KM, Paterson JW: An in vitro study of various drugs on central and peripheral airways of the rat: A comparison with human airways. Br J Pharmacol 1980;68:633–636.

68 Farmer JB, Richards IM, Sheard P, Woods AM: Mediators of passive lung anaphylaxis in the rat. Br J Pharmacol 1975;55:57–64.

69 Sadavongvivad C: Pharmacological significance of biogenic amines in the lung: 5-Hydroxytryptamine. Br J Pharmacol 1970;38:353–365.

70 Pauwels R, Van Der Straeten M, Weyne J, Bazin H: An animal model for the study of the relation between non-specific bronchial reactivity and immunological hypersensitivity. Agents Actions 1983; 13:55–65.

71 Pauwels R, Van Der Strateten M, Weyne J, Bazin H: Genetic factors in non-specific bronchial reactivity in rats. Eur J Respir Dis 1985;66:98–104.

72 Levitt RC, Mitzner WL: Autosomal recessive inheritance of airway hyperreactivity to 5-hydroxytryptamine. J Appl Physiol 1989;67:1125–1132.

73 Martin TR, Cohen ML, Drazen JM: Serotonin-induced pulmonary responses are mediated by the 5-HT$_2$ receptor in the mouse. J Pharmacol Exp Ther 1994;268:104–109.

74 Fryer AD, El-Fakahany EE: Identification of three muscarinic receptor subtypes in rat lung using binding studies with selective antagonists. Life Sci 1990;47:611–618.

75 Wall SJ, Yasuda RP, Li M, Wolfe BB: Development of an antiserum against m$_3$ muscarinic receptors: Distribution of m$_3$ receptors in rat tissues and clonal cell lines. Mol Pharmacol 1991;40:783–789.

76 Eidelman DH, Dimaria GU, Bellofiore S, Wang NS, Guttmann RD, Martin JG: Strain-related differences in airway smooth muscle and airway responsiveness in the rat. Am Rev Respir Dis 1991; 144:792–796.

77 Pauwels RA, Kips JC, Peleman RA, Van Der Straeten M: The effect of endotoxin inhalation on airway responsiveness and cellular influx in rats. Am Rev Respir Dis 1990;141:540–545.

78 Holme G, Piechuta H: The derivation of an inbred line of rats which develop asthma-like symptoms following challenge with aerosolized antigen. Immunology 1981;42:19–24.

79 Brunet G, Piechuta H, Hamel R, Holme G, Ford-Hutchinson AW: Respiratory responses to leukotrienes and biogenic amines in normal and hyperreactive rats. J Immunol 1983;131:434–438.

80 Hamel R, Ford-Hutchinson AW: Pulmonary and cardiovascular changes in hyperreactive rats from citric acid aerosols. J Appl Physiol 1985;59:354–359.

81 Martin TR, Gerard NP, Galli SJ, Drazen JM: Pulmonary responses to bronchoconstrictor agonists in the mouse. J Apl Physiol 1988;64:2318–2323.

82 Weinmann CG, Black CM, Levitt RC, Hirshman CA: In vitro tracheal responses from mice chosen for in vivo lung cholinergic sensitivity. J Appl Physiol 1990;69:274–280.

83 Levitt RC, Mitzner W: Expression of airway hyperreactivity to acetylcholine as a simple autosomal recessive trait in mice. FASEB J 1988;2:2605–2608.

84 Longphre M, Kleeberger SR: Susceptibility to PAF-induced airways hyperreactivity and hyperpermeability: Inter-strain variation and genetic control. Am J Respir Cell Mol Biol 1996;14:461–469.

85 Goldie RG, Paterson JW, Spina D, Wale J: Classification of β-adrenoceptors in human isolated bronchus. Br J Pharmacol 1984;81:611–615.

86 Zaagsma J, Van der Heijden PJ, Schaar MW, Bank CM: Differentiation of functional adrenoceptors in human and guinea-pig airways. Eur J Respir Dis 1986;65(suppl 135):161–163.

87 Henry PJ, Rigby PJ, Goldie RG: Distribution of β$_1$- and β$_2$-adrenoceptors in mouse trachea and lung: A quantitative autoradiographic study. Br J Pharmacol 1990;99:136–144.

88 Ewart SL, Mitzner W, DiSilvestre DA, Meyers DA, Levitt RC: Airway hyperresponsiveness to acetylcholine: Segregation analysis and evidence for linkage to murine chromosome 6. Am J Respir Cell Mol Biol 1996;14:487–495.

89 Ewart SL, Levitt RC, Mitzner W: Respiratory system mechanics in mice measured by end-inflation occlusion. J Appl. Physiol 1995;79:560–566.

90 De Sanctis GT, Merchant M, Beier DR, Dredge RD, Grobholz JK, Martin TR, Lander ES, Drazen JM: Quantitative locus analysis of airway hyperresponsiveness in A/J and C57BL/6J mice. Nat Genet 1995;11:150–154.

Charles W. Emala, MD, The Johns Hopkins School of Hygiene and Public Health,
Division of Physiology/Room 7006, 615 N. Wolfe Street, Baltimore, MD 21205 (USA)

Hall, IP (ed): Genetics of Asthma and Atopy.
Monogr Allergy. Basel, Karger, 1996, vol 33, pp 53–70

....

Human Leukocyte Antigen Genes and Allergic Disease

W. Martin Howell [a], *Stephen T. Holgate* [b]

[a] Molecular Immunology Group and
[b] Immunopharmacology Group, University Medicine, Southampton General
Hospital, Southampton, UK

Introduction

The underlying cause of asthma, hay fever and eczema in children and young adults is atopy, a disorder characterized by persistent immunoglobulin E (IgE) responses to extrinsic protein allergens. The phenotype is comprised of many factors, including total and allergen-specific IgE titres, skin-prick test positivity, medical history and provocation tests of bronchial reactivity. A broad definition of atopy may include 40% of the population [1], while up to 10% of the population of western European countries suffers from asthma. In addition, age affects the penetrance of atopy, with a symptomatic peak and highest IgE levels in the teens, followed by a decline that results in the total serum IgE at age 45 being half that at 15 years [2]. Thus it is hardly surprising that genetic analysis of atopic asthma has been difficult, despite strong evidence for heritability.

Currently, worldwide searches are underway for genetic loci regulating the IgE response and particular interest has centred on the β-subunit of the high-affinity IgE receptor (FCεRIβ) located on chromosome 11q and expressed on mast cells and basophils and more recently shown to be present on dendritic cells and eosinophils [3–10]. While claims have been made that FCεRIβ is a major gene for atopy, with a strong bias towards maternal inheritance [1, 8], these findings have not been confirmed by all other studies [e.g. 7, 10]. More recently, claims have also beeen made that interleukin-4 (IL-4) or a closely linked gene on chromosome 5q regulates IgE production in a nonantigen-specific fashion [11] and in separate studies is linked with airway hyperresponsiveness [12]. These exciting findings are currently under investigation in many laboratories and are considered elsewhere in this volume. However, in addition to genes that control the overall level of expression of the atopic phenotype, there are important genetic influences exerted over the recognition of allergenic antigens and the subsequent propagation of the immune response involving T cells, B cells and IgE synthesis. In particular, genetic epidemiological studies in atopic subjects have demonstrated several significant associations between particular human leukocyte antigen (HLA) class II genotypes and specific allergic IgE responses to common airborne allergens. The majority of these studies have been single-centre analyses of unrelated individuals, but a concerted international collaborative study of the role of HLA genes in a range of allergic IgE responses was mounted in the Eleventh International Histocompatibility Workshop (IHW), with a similar collaborative study of the influence of HLA genes on the development of house dust mite (HDM) allergy in family-based analyses nearing completion in the Twelfth IHW. In this chapter, results from both single-centre and collaborative studies of the role of HLA genes in specific allergies and in the atopic immune response in both population-based and family material will be summarized and reviewed.

Immunogenetics of the HLA System

Classical HLA Antigens

The classical human major histocompatibility complex (MHC) or HLA molecules are encoded by two highly polymorphic gene families located in a 3,600-kb region of chromosome 6p (6p21.3) (fig. 1). The resulting HLA molecules – the most polymorphic found in humans – are membrane-bound glycoproteins that bind processed antigenic peptides and present them to T cells. The HLA class I A, B and C molecules are each composed of an MHC-encoded heavy chain (MW 45 kD), noncovalently associated with a nonpolymorphic polypeptide, β_2-microglobulin (MW 12 kD), encoded on chromosome 15 (fig. 2). There are now known to be 56 different expressed HLA-A alleles, 111 HLA-B alleles and 34 HLA-C alleles, excluding silent substitutions and null alleles [13]. These class I antigens are expressed on all nucleated cells (except fetal trophoblast cells) and platelets and function to present peptides of largely endogenous (viral) origin to CD8 + T cells, which in the main are of cytotoxic function. The bound peptides are highly circumscribed in length, usually 8–9 amino acids, and are held in a peptide-binding groove, which X-ray crystallography has shown to have an allele-specific conformation [reviewed in 14]. The polymorphic residues which distinguish between the different alleles

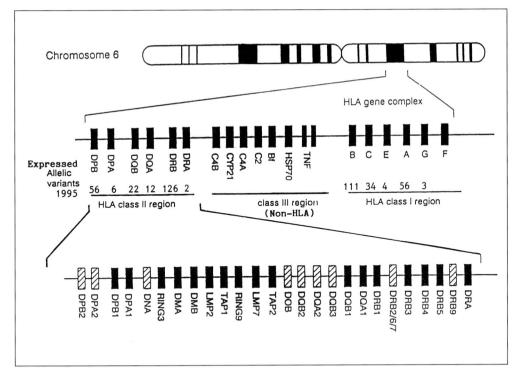

Fig. 1. The HLA gene complex on the short arm of chromosome 6. Loci encompassing genes whose products are known to be expressed are marked with filled squares (derived and updated from Thorsby and Ronningen [31], with numbers of expressed alleles taken from Bodmer et al. [13]).

of a particular HLA class I locus are found, in the main, within this peptide-binding groove [15].

In contrast to class I molecules, HLA class II molecules, comprising three main subclasses – DR, DQ and DP – are found on a more restricted range of cell types, including B cells, activated T cells, the monocyte/macrophage lineage and are also interferon-γ inducible. An expressed class II molecule consists of an α chain (MW 31–34 kD) encoded by an A gene, noncovalently associated with a β chain (MW 26–29 kD), encoded by a B gene (fig. 2). Each DR, DQ or DP subregion consists of at least one expressed A and one expressed B gene (see fig. 1). Both A and B genes may be polymorphic, but most polymorphism resides in the B genes. There are now known to be 2 DRA, 126 DRB, 12 DQA, 22 DQB, 6 DPA and 56 different expressed DPB

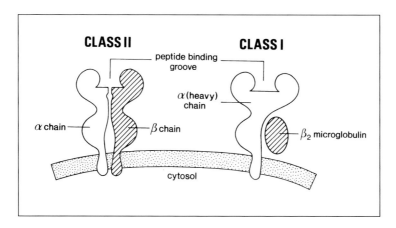

Fig. 2. Basic structure of HLA class I and class II molecules.

alleles, excluding silent substitutions [13]. Both α and β chains combine to form a peptide-binding groove, shown by X-ray crystallography to be very similar to the class I groove [16]. However, class II molecules present peptides of largely exogenous origin to CD4+ T cells of largely 'helper' phenotype. These bound peptides are generally longer and more variable in length than peptides bound to class I molecules (i.e. 14–21 amino acids), due to the more open ends of the peptide-binding groove.

Both classes of HLA molecule function to present self antigens in the thymus and so induce tolerance, while foreign antigens are presented in the context of self HLA molecules in the periphery, invoking an immune response.

Non-classical HLA and Non-HLA Genes in the HLA Class I/II Regions

The application of molecular techniques of cloning, sequencing and gene mapping have also revealed a number of additional HLA and non-HLA genes in the class I/II regions. In the class I region, there are known to be 17 'nonclassical' genes or gene fragments, although only 3 of these – HLA-E, F and G – are known to be transcribed [17] (fig. 1). Little is yet known of the possible functions of HLA-E and F. More is known of HLA-G, which is closely homologous to other class I gene sequences and was thought to show little polymorphism, although this may not be so [18]. HLA-G is primarily although not exclusively expressed on fetal cytotrophoblast cells. These are the only fetally derived cells in contact with maternal cells and lack expression of classical class I genes. In consequence, it is thought that the HLA-G gene product may function as a fetal antigen presenting/recognition molecule and hence in the absence of classical, highly polymorphic class I molecules, may permit maternal tolerance of the placenta [19].

A series of gene mapping studies carried out independently in the laboratories of John Trowsdale (ICRF, London) and Thomas Spies (Harvard), plus similar studies of the rat MHC by John Monaco (Virginia), have revealed a series of novel genes in the class II region, located between the DQ and DP subregions [20–24]. Gene sequencing, deletion mutant and transfection studies have now demonstrated a role for many of these genes in pathways of antigen processing and presentation. While HLA class I and II molecules are synthesized and assembled in the endoplasmic reticulum and peptide binding to class I molecules also occurs here, it has been a conundrum as to how these peptides are generated from proteins present in the cytosol and transported into the endoplasmic reticulum. It is now known that the proteasome, a cytoplasmic complex of at least 16 polypeptides (each of MW 15–30 kD), catalyses the degradation of the vast majority of cell proteins and generates most peptides presented by class I molecules [25]. Two subunits of the proteasome are encoded by two genes locted between DQ and DP – LMP2 and LMP7 (LMP = large multifunctional protease) (fig. 1). Deletion of these LMP2/LMP7 genes alters the nature of the peptides generated by the proteasome, so that they no longer have optimal characteristics for class I binding [26]. Two additional genes, TAP1 and TAP2 (TAP = transporter of antigen peptides) also located in the DQ-DP interval (fig. 1) encode separate chains of a trans-endoplasmic reticulum membrane heterodimer which functions as a peptide pump, transporting peptides generated by the proteasome into the endoplasmic reticulum. The TAP genes show some polymorphism and this may influence the nature of the peptides transported, with results suggesting that the TAP transporter molecule preselects peptides according to sequence and length in a manner compatible with subsequent presentation by class I molecules [27].

The DQ-DP region is still richer in what were once termed RING or 'really interesting new genes', as defined by John Trowsdale's group. Two further genes – DMA and DMB – map to this region (see fig. 1) and have sequences intermediate between those of classical class I and II genes, but may encode a class II-like heterodimer with a modified (more rigid) peptide-binding groove [28]. Recent transfection experiments in mutant B lymphoblastoid cell lines suggest that HLA-DM is expressed and appears to function at an intracellular site to promote peptide binding to classical class II molecules. Peptide binding to class II molecules in the endoplasmic reticulum is prevented by co-assembly of the α and β chains, with a third chain, the so-called invariant chain (Ii, MW 35 kD, encoded by a gene on chromosome 5). The Ii chain also acts as an 'address label', directing the class II-Li complex to an intracellular endosomal compartment [29]. It is currently thought that HLA-DM acts as a 'sink' for the removal of Ii chain-derived 'CLIP' (class II-associated invariant chain peptides) peptides in this compartment, so freeing classical class II molecules for peptide binding [reviewed in 30].

Taken together, these collected discoveries have overthrown earlier concepts of the MHC class I and II regions as solely containing genes encoding for molecules which present antigenic peptides to T cells. Rather, the current view is of a genetic region encoding many different types of molecule collectively involved in pathways of antigen processing and presentation to helper and cytotoxic T cells. All of these gene products may have a role in immunologically mediated, HLA-associated diseases, although most studies have been directed towards the 'classical' HLA class I and more particularly, class II genes. Nevertheless, investigation of the possible role of polymorphism of certain 'nonclassical' MHC genes, such as TAP1 and TAP2, in HLA-associated diseases has recently been of considerable interest [e.g. 32, 33].

HLA and Disease

Due to the pivotal role of HLA molecules in regulating the T-cell immune response, combined with the exuberant polymorphism of these loci, it is not surprising that particular HLA alleles (especially of the class II loci) have been implicated in susceptibility to a wide range of diseases with an immunological basis. Since the first reported association between a particular HLA class I allele and predisposition to Hodgkin's disease in 1967 [34], more than 500 conditions have been found to be associated with one or more genes in the HLA complex, including rheumatoid arthritis, multiple sclerosis and type 1 insulin-dependent diabetes mellitus [35]. For many of these diseases,the HLA associations are weak and may be fortuitous. For other diseases, the HLA associations are so strong that they are almost certainly the result of a direct involvement of the HLA genes concerned in the pathogenesis of the disease, e.g. HLA-DQA1*0501, DQB1*0201 and coeliac disease [36] (the World Health Organization HLA nomenclature committee now specifies a 4 or 5 number code to define each allele of each HLA locus at the DNA level, with 2 number codes defining the previous lower resolution serological equivalents). While most studies of the influence of HLA polymorphism on disease susceptibility have concentrated upon autoimmune and to a lesser extent malignant diseases, a number of studies from the late 1970s onwards have examined the relationship between HLA and atopy, allergy and asthma. Results from the major single and multi-centre analyses of unrelated and familial study groups are summarized in the following sections.

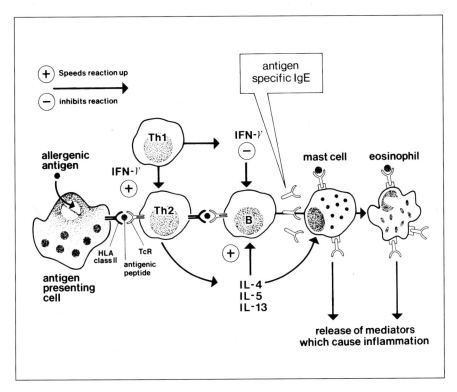

Fig. 3. From antigen to allergic response, showing interaction of HLA class II plus antigenic peptide and TcR on T$_h$-2 helper T cells. IFN-γ = Interferon-γ; IL = interleukin.

Population-Based Studies of the Role of HLA Class II Polymorphism in Atopic Allergy

Single-Centre Studies

The most probable role of the interaction between the HLA class II-allergenic peptide complex and the T-cell receptor (TCR) in setting up the IgE-mediated allergic response is shown in figure 3. In theory, protein antigens have multiple epitopes or sequence motifs that can each be presented in association with different HLA class II polymorphic molecules expressed on the surface of antigen-presenting cells. This phenomenon inevitably dilutes any association between particular HLA class II alleles and specific IgE responses to a given allergen in a population, even when considering purified allergens rather than crude allergen extracts. Despite this, a number of HLA class II-allergen-specific IgE response associations have been substantiated by single-centre, population-based studies.

The first studies of possible associations between HLA antigens and asthma and allergic responses to crude pollen and HDM allergen were performed during the late 1970s, but results were variable [37–39] in part due to the relatively poor definition of the HLA loci and their allelic polymorphism in the pre-DNA era. More systematic early studies of the possible role of HLA polymorphism in atopic allergy, dating from the early 1980s onwards, concentrated on pollen allergens and clearly showed that increased IgE production to ragweed (*Antemisia artemisiifolia*) pollen allergens Amb a 5, Amb t 5, Amb p 5 and Amb a 6 occurs in individuals expressing DRB1*1501 (DR2)-associated haplotypes [40–43]. A more recent study of unrelated individuals has suggested a more complex picture, in which HLA-DR2 or a DR2-associated gene or genes, specifies high levels of IgE anti-Amb a 5 and is associated with allergic asthma but not rhinitis in patients with ragweed atopy. Conversely, in this latter study the DR3-associated 'extended haplotype' was associated with rhinitis only, as compared with general controls [44]. All of these studies, although reporting convincing associations, relied upon serological rather than the more recently developed accurate and high-resolution DNA-based methods for detection of HLA class II polymorphism.

Similar single-centre studies have established population associations between DRB1*03 and DRB1*11 alleles and immune responses to rye grass (*Lolium perenne*) antigens [45, 46]. In addition, it has been established that products of the HLA-DRB1, DRB3 and DRB5 genes are able to restrict the recognition of HDM (*Dermatophagoides* spp.) determinants [47, 48] in line with population associations demonstrated in the Eleventh IHW [49] (see below).

Several other single-centre studies have provided evidence for a number of additional IgE associations between particular HLA class II alleles and increased IgE response to specific purified allergens. Notable among these are positive associations between DRB1*07 and DQB1*0201 and IgE response to the main olive pollen allergen Ole e 1 [50], DRB3*0101 and IgE responsiveness to the birch pollen allergen Bet v 1 [51], and DRB1*0101, DQA1*0101, DQB1*0501 and IgE responsiveness to the chironomid nonbiting midge allergen Chi t 1 [52]. Furthermore, it has been suggested that the HLA-DR or DQ loci may play a protective role in controlling the IgE response to mellitin, a component of bee venom [53], while a component of the T-cell repertoire reactive with ovalbumin in hen's egg-allergic patients may be restricted by HLA-DP, with low immune responsiveness in hen's egg-sensitive patients with atopic dermatitis regulated by HLA-DQ [54].

These single-centre studies of unrelated individuals have revealed a number of significant associations between particular HLA class II alleles, especially those of the HLA-DR loci, and IgE response to various purified allergies, while other workers have demonstrated a role for other HLA class II loci, including DQ and DP, in regulation of the allergen-specific immune response. Nevertheless, the most comprehensive investigation of HLA and allergy performed to date is still that of the eleventh IHW [49].

The Eleventh IHW

In the Eleventh IHW, specific IgE (and IgG where measured) responsiveness to 8 highly purified allergens (Lol p 5, Amb a 5, Par o 1, Bet v 1, Cry i 1, Fel d 1, Alt a 1, and Der p 1) was examined in 1,006 atopic individuals in 13 population groups, with the main emphasis on correlations with polymorphism of the HLA-DRB1 gene (the most polymorphic HLA class II gene). The most consistent associations found were between DRB1*04 and 14 and IgE antibody responsiveness to the *Alternaria* mould allergen Alt a 1, with the DRB1*04 association found across racial and ethnic lines, while the DRB1*14 association was found in 3 Caucasian population groups, despite the relative rareness of Alt a 1 responsiveness. The highest rate of antibody responsiveness was towards Fel d 1 (the major cat allergen), with DRB1*15 showing a positive association with Fel d 1 IgE antibodies in two study groups (one European, one Japanese) and a negative association in a third group (also Japanese). Other associations included confirmation of the well-known association between IgE response to the Amb a 5 allergen and DRB1*15 (in US Caucasians), positive associations between DRB1*04 (in Swedish Caucasians) and DRB1*03 (in Italian Caucasians) and IgE responsiveness to the HDM allergen Der p 1, plus a negative association between DRB1*11 and IgE antibodies to Bet v 1 in Swedish and Bulgarian populations.

Thus the Eleventh IHW and similar, more limited studies have shown that while there may still be considerable uncertainty over the exact nature of particular HLA class II associations with specific IgE responsiveness to defined allergens (the DRB1*15 association with antibody response to Amb a 5 is still the only consistent association seen in several studies), such associations do indeed occur. The goal of future studies is to definitively establish the exact nature and relative significance of these HLA genetic contributions to specific IgE responses to allergen and then to determine the relative genetic contribution to both specific and total IgE responses in allergy. To achieve the first of these aims, two new approaches will be required. Firstly, it will be necessary to move from largely population-based studies of unrelated atopic individuals to family-based analyses, which permit additional valuable sibpair tests for genetic linkage to be performed. Secondly, the advent of comprehensive DNA-based techniques for HLA class II DR, DQ and DP typing should now permit full molecular characterization of segregating HLA class II haplotypes in the families concerned [55].

Family-Based Studies of the Role of HLA Class II Polymorphism in Atopic Allergy

A Large Single-Centre Study and Comparisons with the Eleventh IHW
One recent family-based study shows the value of comprehensive HLA class II DNA typing for full characterization of segregating HLA class II haplotypes in the families concerned [56]. In this study of 431 British Caucasian subjects in 77 nuclear and 7 extended families recruited through allergy or asthma clinics, HLA-DRB1 and DPB1 genotype frequencies were determined by polymerase chain reaction (PCR)-based DNA typing and specific IgE responses to Der p 1 and Der p 2 (from the HDM), Alt a 1, Can f 1 (from the dog), Fel d 1 and Phl p 5 (from the timothy grass, *Phleum pratense*) were determined. Results demonstrated a number of weak associations between HLA-DRB1 alleles and IgE responses to the test allergens, of which the strongest association was between DRB1*01 and Fel d 1 responsiveness. This association was also supported by excess sharing of the HLA haplotype concerned in affected sibling pairs, as was the association between IgE responsiveness to Alt a 1 and DRB1*04, demonstrating the advantage of studying multicase families in HLA and disease association studies. Logistical regression analyses to control for reactions to more than one allergen also showed that an apparent association between IgE response to Phl p 5 and DRB1*04 was due to the presence of many individuals showing an IgE response to both Alt a 1 and Phl p 5.

It is informative to set the results from this large, single-centre family-based study alongside those of the Eleventh IHW, since three allergens were common to both studies (Der p 1, Alt a 1 and Fel d 1). Firstly, while the association between DRB1*01 and IgE response to Fel d 1 was the strongest association seen in the single-centre British study, no associations were seen in the Eleventh IHW, which did not include a British patient cohort. This difference may be due to differing levels of exposure to the sensitizing allergen in the populations studied, perhaps reflecting British fondness for the domestic cat! Secondly, an association between Alt a 1 IgE responsiveness and DRB1*04 was observed. In the IHW study, similar HLA-DRB1*04 associations were seen in unrelated Bulgarian and Israeli atopic subjects. Conversely, associations with DRB1*14 were seen in US Caucasoid and Swedish individuals. These apparently conflicting findings may be elucidated by further studies of complete

HLA class II haplotypes in these populations. For example, the same DQB1 alleles can be found on some DRB1*04 and DRB1*14 haplotypes and thus the most precise association may in fact be with a DQA1 or DQB1 allele or allele combination. Finally, in the British study, no associations were found between IgE response to Der p 1 and any DRB1 or DPB1 allele. However, in the Eleventh IHW, associations with DRB1*04 and DRB1*03 were seen in Swedish and Italian subjects respectively, as outlined above, although no associations were seen in US Caucasoid subjects. HLA-DPB1 associations were not examined in the Eleventh IHW, but the possible role of DPB1 alleles in IgE responsiveness to Der p 1 has been a matter of debate since a negative association between DPB1*0401 and allergic asthma was reported in a Colombian mulatto study population [57]. This association has been substantiated by the finding that a component of the T-cell repertoire reactive with *Dermatophagoides* spp. determinants in atopic individuals is HLA-DP restricted [58], while DPB1*0401 is negatively associated with both aspirin-tolerant and -intolerant asthma in German and British subjects [59].

HDM Allergy and the Twelfth IHW

The role of particular HLA class II alleles in IgE responsiveness to Der p 1 determinants and the identification of HLA-associated Der p 1 epitopes is currently the object of much study in several laboratories worldwide. In addition, the primary aim of the Twelfth IHW now nearing completion is to undertake a systematic analysis of the HLA immunogenetics of atopic HDM allergy using family-based studies in several population groups. This proposed workshop is thus a worthy successor to the Eleventh IHW, building on its strengths and the valuable network of clinical and laboratory teams established, but with the added benefits of being more focused, using family rather than population studies and making full use of accurate, high-resolution DNA techniques for HLA class II gene polymorphism detection. The aim of the study, coordinated by Prof. Malcolm Blumenthal of the University of Minnesota, is the collection of at least 25 nuclear families from each participating centre, with the proband and at least 1 sibling having mite sensitivity and only 1 or none of the parents being atopic. Full health questionnaires have been completed for all individuals participating and core investigations include skinprick test reactivity using standard methodology (for at least Der p, Der f, Fel d and Lol p), total and Der p 1, Der f 1, Fel d 1 and Lol p 1 specific IgE and IgG determinations and comprehensive DNA-based HLA class II typing. Complete HLA class II haplotype characterization in affected and unaffected individuals will permit the most precise association between HLA class II marker and specific IgE response to be identified, thus confounding the effects of linkage disequilibrium (preferential associations between alleles of different loci), which is such a frustrating feature of HLA-disease studies. These studies also include analysis of three microsatellites within the HLA region, including the tumor necrosis factor (TNF) loci [60] and the DQA1/DQB1 region [61]. Additional studies running in parallel with the 'core' Twelfth IHW study will focus on studies of T-cell markers (including TCR variable (V) gene repertoire) and polymorphism of other 'candidate' genetic markers, including those genes already implicated on chromosome 5q and 11q. The workshop will be completed by June 1996 and this comprehensive approach should clearly define the precise nature of and relative significance of the HLA class II genetic contribution to HDM allergy in a number of diverse ethnic groups. By combining studies of HLA class II genes and other 'candidate' genetic markers, it will also be possible to dissect the relative genetic contributions to both total and specific IgE responses in this particular allergy.

HLA Class II Genes and Industrial Asthma

Until recently it has not been clear whether occupational asthma, triggered by exposure to small molecules, has a similar immunological basis to that demonstrable in extrinsic allergic asthma. A priori, it is quite likely that a T-cell immune response would be directed towards 'neoantigens' formed by the reaction between the inhaled small molecule in question and particular body proteins. A number of recent studies of HLA class II associations in specific occupational asthmas has suggested that such immunological mechanisms may indeed be involved.

One such study of unrelated European Caucasoid individuals with isocyanate-induced asthma (IAA), compared with ethnically similar healthy control individuals, exposed to the same environment, has demonstrated that allele DQB1*0503 and the DQB1*0201/0301 combination were associated with susceptibility to the disease, while DQB1*0501 and the DQA1*0101, DQB1*0501, DRB1*01 haplotype conferred significant protection [62, 63]. These results are consistent with the hypothesis that immune mechanisms are involved in IAA asthma and that specific genetic factors, including HLA or HLA-linked genes, may increase the risk of developing IAA in exposed workers. These results are highly significant because isocyanates – highly reactive, low-molecular-weight chemicals – are widely used in industry and are the most common cause of occupational asthmas. A similar study has also reported an association between HLA-DRB1*03 and IgE response to acid anhydrides among acid anhydride workers in the paint and varnish industry, again suggesting an HLA-modulated immunological mechanism for another form of occupational asthma involving an inhaled agent [64].

HLA Class II Genes, Allergic Disease and Atopy: Selection of Appropriate Controls for Association Studies

Most studies of the role of HLA genes in allergic disease have concentrated on investigations of the role of the extensive HLA class II polymorphism in regulating antigen-specific IgE responsiveness, with definite but variable results and generally have not suggested a role for class II genes in predisposition to atopy. Naturally, the success of such HLA association studies critically depends on accurate definitions of clinical phenotype, because different clinical subsets of the same disease may vary in their associations with HLA class II alleles and misclassification of subjects may therefore be an important confounding factor. However, the selection of appropriate controls for such studies may be equally important. The apparent lack of an association between HLA class II alleles and atopy, specific IgE response or asthma in a given study may be due to the inclusion of an inappropriate control group. For example, unselected control individuals, drawn from the general population, may comprise up to 40% atopic individuals, confounding analyses of genetic influences on atopy and allergic asthma. Thus, for a definitive study, stringent criteria should be applied to select for nonatopic control subjects.

Using such a rigorous approach, one study of an extended 33 member French family (19 asymptomatic, 14 asthmatic, atopic or both) has reported that the DRB4 (DRB1*04/07) haplotype is associated with both allergic asthma *and* atopy [65]. However, no such DRB4 association with asthma or atopy was found in a study of 22 British HDM-allergic families performed in our own laboratory [66]. Another study from our own centre, of a series of 21 two- and three-generation asthma families, demonstrated a significant association between the DRB1*08, DQB1*04 haplotype and nonatopic status, suggesting that DRB1*08 (or a linked marker) protects against the development of atopy [67].

To add to an as yet inconclusive picture, a third study of 20 two- and three-generation families assertained though a proband as having asthma, failed to find evidence for linkage between atopy or bronchial responsiveness and the HLA region or chromosome 6p, via lack of linkage with D6S105, a polymorphic marker 7 cM from HLA-DR [68]. Thus, despite careful selection of nonatopic controls in familial studies, results are still conflicting. These results eloquently underline the difficulty inherent in genetic studies of a variable and variably penetrant trait such as atopy, but do not rule out a modulating role for HLA class II polymorphism in the regulation of total IgE responsiveness and associated atopic traits in the general population.

Immunological Basis of HLA Class II Allelic Associations with Specific Allergic Responses

Given that a number of positive and negative associations between particular HLA class II alleles and specific IgE responses to allergens have now been established by fairly extensive studies and are being characterized more accurately in ongoing studies, by what immunological mechanisms may these HLA class II genetic associations with atopic allergy be operating? Positive associations between specific IgE responsiveness and HLA class II alleles are consistent with the concept of strong CD4+ helper T cell (Th-2) responses to immunodominant regions of processed class II-presented allergenic antigen, regulating IgE-mediated hypersensitive responses (fig. 3). This model is supported by studies such as that of Higgins et al. [69], who demonstrated that in a single HDM-allergic individual, the T-cell response to *Der p* I was limited to a single region of the protein. By mapping the fine antigen specificity with T-cell clones, these workers identified a cluster of at least three overlapping T-cell epitopes, one of which (residues 110–131) was restricted by DRB1*0101, a second (residues 110–119) by DPB1*0402 and a third epitope (residues 107–119) was presented by DPB1*0401, DPB1*0402 and DPB1*0501, illustrating the heterogeneity of the HLA restriction specificity of this region of *Der p* I. Conversely, negative HLA class II associations may reflect weak processed antigen presentation by the expressed class II molecule in question, lack of T-cell recognition or responsiveness, or more controversially, recognition by 'suppressor' T cells. However, a third possible mechanism should not be ruled out, namely that the HLA haplotype of an individual influences the expressed T-cell repertoire of that individual during thymic development, thus selecting for or deleting potential HLA plus allergen peptide-reactive T cells. In fact, there is now an increasing body of evidence that the T-cell repertoire (as measured by TCR β chain V gene family usage) of an individual is indeed influenced by HLA haplotype, while skewing of the expressed TcR repertoire within the CD4 and CD8 T-cell subsets may result from interaction of these subsets with HLA class II and I molecules respectively during development [70]. However, the degree and significance of these HLA-mediated effects on TcR repertoire in T-cell-mediated disease remains controversial.

In summary: While there remains considerable uncertainty as to the exact role of particular HLA class II genetic polymorphisms in triggering an allergen-specific IgE response, and the relative importance of specific versus total IgE responses in allergic disease, there can be little doubt that the HLA genes do play a crucial role. The more systematic studies of the relationship between HLA and other loci implicated in regulating the atopic response currently underway, including that of the Twelfth IHW, should provide further clarification of this. These studies will also address the possible role of polymorphism associated with the MHC class III TNF α and β genes in the development of atopic disease. However, as yet, the influence (if any) of non-HLA genes within the human MHC, including the TAP and LMP genes, on the generation of allergen-specific T-cell-mediated responses remains unknown. The ongoing in vitro characterization of further allergenic T-cell epitopes and the determination of their HLA restriction specificity will not only provide a theoretical basis for epidemiological observations of HLA class II allergen-specific IgE response associations, but may ultimately provide information for the design of peptide-based immunotherapy, targetted at CD4 + T cells, in the management of specific allergic responses.

References

1   Cookson WOCM, Hopkin JM: Dominant inheritance of atopic immunoglobulin E responsiveness. Lancet 1988;i:86–88.

2   Cline MG, Burrows BB: Distribution of allergy in a population sample residing in Tucson, Arizona. Thorax 1989;44:425–431.

3   Cookson WOCM, Sharp PA, Faux JA, Hopkin JM: Linkage between immunoglobulin E responses underlying asthma and rhinitis and chromosome 11q. Lancet 1989;i:1292–1295.

4   Morton NE: Major loci for atopy? Clin Exp Allergy 1992;22:1041–1043.

5   Marsh DG, Meyers DA: A major gene for allergy – Fact or fancy? Nat Genet 1992;2:252–254.

6   Young RP, Sharp PA, Lynch JR, Faux JA, Lathrop GM, Cookson WOCM, Hopkin JM: Confirmation of genetic linkage between atopic IgE responses and chromosome 11q13. J Med Genet 1992; 29:236–238.

7   Moffat MR, Sharp PA, Faux JA, Young RP, Cookson WOCM, Hopkin JM: Factors confounding genetic linkage between atopy and chromosome 11q. Clin Exp Allergy 1992;22:1046–1051.

8   Cookson WOCM, Young RP, Sandford AJ, Moffat MF, Shirakawa T, Sharp PA: Maternal inheritance of atopic IgE responsiveness on chromosome 11q. Lancet 1992;340:381–384.

9   Shirakawa T, Li A, Duboritz M, Dekker JW, Shaw AE, Faux JA, Ra C, Cookson WOCM, Hopkin JM: Association between atopy and variants of the β subunit of the high-affinity immunoglobulin E receptor. Nat Genet 1994;7:125–130.

10 Brereton HM, Ruffin RE, Thompson PJ, Turner DR: Familial atopy in Australian pedigrees: Adventitious linkage to chromosome 8 is not confirmed, nor is there evidence of linkage to the high affinity IgE receptor. Clin Exp Allergy 1994;24:868–877.

11 Marsh DG, Neely JD, Breazeale DR, Ghosh B, Freidhoff LR, Ehrlich-Kautzky E, Shou C, Krishnaswamy G, Beaty TH: Linkage analysis of IL-4 and other chromosome 5q 31.1 markers and total serum immunoglobulin E concentrations. Science 1994;264:1152–1156.

12 Postma DS, Bleecker ER, Amelung PJ, Holroyd KJ, Xu JF, Panhuysen CIM, Meyers DA, Levitt RC: Genetic susceptibility to asthma – Bronchial hyperresponsiveness coinherited with a major gene for atopy. N Engl J Med 1995;333:894–900.

13 Bodmer JG, Marsh SGE, Albert ED, Bodmer WF, Dupont B, Erlich HA, Mach B, Mayr WR, Parham P, Sasazuki T, Schreuder GMTh, Strominger JL, Svejgaard A, Terasaki PI: Nomenclature for factors of the HLA system. Hum Immunol 1995;43:149–164.

14 Stern LJ, Wiley DC: Antigenic peptide binding by class I and class II histocompatibility proteins. Structure 1994;2:245–251.

15 Parham P: Typing for class I HLA polymorphism: Past, present and future. Eur J Immunogenet 1992;19:347–359.

16 Brown JH, Jardetsky TS, Gorga JC, Stern LJ, Urban RG, Strominger JL, Wiley DC: Three-dimensional structure of the human class II histocompatibility antigen HLA-DR1. Nature 1993; 364:33–39.

17 Le Bouteiller P: HLA class I chromosomal region, genes and products: Facts and questions. Crit Rev Immunol 1994;14:89–129.

18 Van der Ven K, Ober C: HLA-G polymorphism in African Americans. J Immunol 1994;153: 5628–5633.

19 Schmidt CM, Orr HT: Maternal fetal interactions: The role of the MHC class I molecule HLA-G. Crit Rev Immunol 1993;13:207–224.

20 Trowsdale J, Hanson I, Mockridge I, Beck S, Townsend A, Kelly A: Sequences encoded in the class II region of the MHC related to the 'ABC' superfamily of transporters. Nature 1990;348:741–744.

21 Glynne R, Powis SH, Beck S, Kelly A, Kerr L-A, Trowsdale J: A proteasome-related gene between the two transporter loci in the class II region of the human MHC. Nature 1991;353:357–360.

22 Spies T, Bresnahan M, Bahram S, Arnold D, Blanck G, Mellins E, Pious D, DeMars R: A gene in the human major histocompatibility complex class II region controlling the class I antigen presentation pathway. Nature 1990;348:744–747.

23 Deverson EV, Gow IR, Coadwell WJ, Monaco JJ, Butcher GW, Howard JC: MHC class II region encoding proteins related to the multidrug resistance family of transmembrane transporters. Nature 1990;348:738–740.

24 Brown MG, Driscoll J, Monaco JJ: Structural and serological similarity of MHC-linked LMP and proteasome (multicatalytic proteinase) complexes. Nature 1991;353:355–357.

25 Rock KL, Gramm C, Rothstein L, Clark K, Stein R, Dick L, Hwang D, Goldberg AL: Inhibitors of the proteasome block the degradation of most cell proteins and the generation of peptides presented on MHC class I molecules. Cell 1994;78:761–771.

26 Driscoll J, Brown M, Finley D, Monaco J: MHC-linked LMP gene products specifically alter peptidase activities of the proteasome. Nature 1993;365:262–264.

27 Momburg F, Neefjes JJ, Hammerling GJ: Peptide selection by MHC-encoded TAP transporters. Curr Opin Immunol 1994;6:32–37.

28 Kelly AP, Monaco JJ, Cho S, Trowsdale J: A new human HLA class II-related locus, DM. Nature 1991;353:571–573.

29 Sant AJ, Miller J: MHC class II antigen processing: Biology of invariant chain. Curr Opin Immunol 1994;6:57–63.

30 Engelhard VH: How cells process antigens. Sci Am 1994;Aug:44–51.

31 Thorsby E, Ronningen KS: Role of HLA genes in predisposition to develop insulin-dependent diabetes mellitus. Ann Med 1992;24:523–531.

32 Marsal S, Hall MA, Panayi G, Lanchbury JS: Association of TAP2 polymorphism with rheumatoid arthritis is secondary to allelic association with HLA-DRB1. Arthritis Rheum 1994;37:504–513.

33 Middleton D, Megaw G, Cullen C, Hawkins S, Darke C, Savage DA: TAP1 and TAP2 polymorphism in multiple sclerosis patients. Hum Immunol 1994;40:131–134.

34 Amiel JL: Study of the leukocyte phenotypes in Hodgkin's disease; in Cutoni ES, Mattiuz PL, Tosi RM (eds): Histocompatibility Testing. Copenhagen, Munksgaard, 1970, pp 79–81.

35 Tiwari JL, Terasaki PI: HLA and Disease Associations, New York, Springer, 1985.

36 Howell WM, Leung ST, Jones DB, Nakshabendi I, Hall MA, Lanchbury JS, Ciclitira PJ, Wright DH: HLA-DRB, -DQA, and -DQB polymorphism in celiac disease and enteropathy-associated T cell lymphoma: Common features and additional risk factors for malignancy. Hum Immunol 1995; 43:29–37.

37 Morris MJ, Vaughan H, Lane DJ, Morris PJ: HLA in asthma. Monogr Allergy. Basel, Karger, 1977, pp 30–34.

38 Perrin-Fayolle M, Betuel H, Biot N, Grosclaude M: HLA antigens in patients allergic to pollen and house dust. Monogr Allergy. Basel, Karger, 1977, p 74.

39 Turton CWG, Morrish L, Buckingham JA, Lawler SD, Turner-Warwick M: Histocompatibility antigens in asthma: Population and family studies. Thorax 1979;34:670–676.

40   Marsh DG, Hsu SH, Roebber M, Ehrlich-Kautzky E, Freidhoff LR, Meyers DA, Pollard MK, Bias WB: HLA-Dw2: A genetic marker for human immune response to short ragweed pollen allergen Ra5. I. Response resulting primarily from antigenic exposure. J Exp Med 1982;155:1439–1451.

41   Roebber M, Klapper DG, Goodfriend L, Bias WB, Hsu SH, Marsh DG: Immunochemical and genetic studies of *Amb t* v (Ra 5 G), an Ra homologue from giant ragweed pollen. J Immunol 1985;134:3062–3069.

42   Marsh DG, Freidhoff LR, Ehrlich-Kautzky E, Bias WB, Roebber ML: Immune responsiveness to *Ambrosia artemisiifolia* (short ragweed) pollen allergen *Amb a* VI (Ra 6) is associated with HLA-DR5 in allergic humans. Immunogenetics 1987;26:230–236.

43   Marsh DG, Zwollo P, Huang SK, Ghosh B, Ansari AA: Molecular studies of human immune response to allergens. Cold Spring Harb Symp Quant Biol 1989;54:459–470.

44   Blumenthal M, Marcus-Bagley D, Awdeh Z, Johnson B, Yunis EJ, Alper CA: HLA-DR2, [HLA-B7, SC31, DR2] and [HLA-B8, SC01, DR3] haplotypes distinguish subjects with asthma from those with rhinitis only in ragweed pollen allergy. J Immunol 1992;148:411–416.

45   Freidhoff LL, Ehrlich-Kautzky EE, Meyers DA, Ansari AA, Bias WB, Marsh DG: Association of HLA-DR3 with human immune response to *Lol p* I and II allergens in allergic subjects. Tissue Antigens 1988;31:211–219.

46   Ansari AA, Friedhoff LR, Meyers DA, Bias WB, Marsh DG: Human immune responsiveness to *Lolium perenne* pollen allergen *Lol p* III (rye III) is associated with HLA-DR3 and DR5. Hum Immunol 1989;25:59–71.

47   O'Hehir RE, Eckels DD, Frew AJ, Kay AB, Lamb JR: MHC-class II restriction specificity of cloned human T-lymphocytes reactive with *Dermatophagoides farinae* (house dust mite). Immunology 1988; 64:627–631.

48   O'Hehir RE, Mach B, Berte C, Greenlaw R, Tiercy J-M, Bal V, Lechler RI, Trowsdale I, Lamb JR: Direct evidence for a functional role of HLA-DRB3 gene products in the recognition of *Dermatophagoides* spp. by helper T-cell clones. Int Immunol 1990;2:885–892.

49   Marsh DG, Blumenthal MN, Ishikawa T, Ruffilli A, Sparholt S, Freidhoff LR: HLA and specific immune responsiveness to allergens; in Tsuji K, Aizawa M, Sasazuki T (eds):HLA 1991. Proc Eleventh International Histocompatibility Workshop and Conference. Oxford, Oxford University Press, 1992, vol I, pp 765–767.

50   Cárdaba B, Vilches C, Martin E, Andrés B de, Pozo V del, Hernández D, Gallardo S, Fernandez JC, Villalba M, Rodriguez R, Basomba A, Kreisler M, Palomino P, Lahoz C: DR7 and DQ2 are positively associated with immunoglobulin-E response to the main antigen of olive pollen (*Ole e* I) in allergic patients. Hum Immunol 1993;38:293–299.

51   Sparholt SH, Georgsen J, Madsen HO, Svendsen UG, Schou C: Association between HLA-DRB3*0101 and immunoglobulin E responsiveness to *Bet v* I. Hum Immunol 1994;39:76–78.

52   Tautz C, Rihs H-P, Thiele A, Zwollo P, Freidhoff LR, Marsh DG, Baur X: Association of class II sequences encoding DR1 and DQ5 specificities with hypersensitivity to chironomid allergen, *Chi t* I. J Allergy Clin Immunol 1994;93:918–925.

53   Lympany P, Kemeny M, Welsh KI, Lee TH: An HLA-associated nonresponsiveness to mellitin: A component of bee venom. J Allergy Clin Immunol 1990;86:160–170.

54   Shinbara M, Kondo N, Agata H, Fukutomi O, Nishida T, Kobayashi Y, Orii T: T cell proliferation restricted by HLA class II molecules in patients with hen's egg allergy. Exp Clin Immunogenet 1995;12:103–110.

55   Kimura A, Sasazuki T: Eleventh International Histocompatibility Workshop reference protocol for the HLA DNA-typing technique; in Tsuji K, Aizawa M, Sasazuki T (eds): HLA 1991. Proc Eleventh International Histocompatibility workshop and Conference, Oxford, Oxford University Press, 1992, vol I, pp 397–419.

56   Young RP, Dekker JW, Wordsworth BP, Schou C, Pile KD, Mathiesen F, Rosenberg WMC, Bell JI, Hopkin JM, Cookson WOCM: HLA-DR and HLA-DP genotypes and immunoglobulin E responses to common major allergens. Clin Exp Allergy 1994;24:431–439.

57   Caraballo L, Marrugo J, Jimenez S, Angelini G, Ferrara GB: Frequency of DPB1*0401 is significantly decreased in patients with allergic asthma in a mulatto population. Hum Immunol 1991;32:157–161.

58 Higgins JA, Lamb JR, Marsh SGE, Tonks S, Hayball JD, Rosen-Bronson S, Bodmer JG, O'Hehir RE: Peptide-induced nonresponsiveness of HLA-DP restricted human T-cells reactive with *Dermatophagoides* spp. (house dust mite). J Allergy Clin Immunol 1992;90:749–756.

59 Lympany PA, Welsh KI, Christie PE, Schmitz-Schumann M, Kemeny M, Lee TH: An analysis with sequence-specific oligonucleotide probes of the association between aspirin-induced asthma and antigens of the HLA system. J Allergy Clin Immunol 1993;92:114–123.

60 Nedospasov SA, Udalova IA, Kuprash DV, Turetskaya RL: DNA sequence polymorphism at the human tumor necrosis factor (TNF) locus: Numerous TNF/lymphotoxin alleles tagged by two closely linked microsatellites in the upstream region of the lymphotoxin (TNF-β) gene. J Immunol 1991;147:1053–1059.

61 Macaubas C, Hallmayer J, Kalil J, Kimura A, Yasunaga S, Grumet FC, Mignot E: Extensive polymorphism of a (CA)n microsatellite located in the HLA-DQA1/DQB1 class II region. Hum Immunol 1995;42:209–220.

62 Bignon JS, Aron Y, Ju LY, Kopferschmitt MC, Garnier R, Mapp C, Fabbri LM, Pauli G, Lockhardt A, Charron D, Swierczewski E: HLA class II alleles in isocyanate-induced asthma. Am J Respir Crit Care Med 1994;149:71–75.

63 Fabbri LM, Mapp CE, Balboni A, Baricordi R: HLA class II molecules and asthma induced by toluene diisocyanate. Int Arch Allergy Immunol 1995;107:400–401.

64 Young RP, Barker RD, Pile KD, Cookson WOCM, Newman Taylor AJ: The assocation of HLA-DR3 with specific IgE to inhaled acid anhydrides. Am J Respir Crit Care Med 1995;151:219–221.

65 Aron Y, Swierczewski E, Lockhart A: A simple and rapid micromethod for genomic DNA extraction from jugal epithelial cells. Allergy 1994;49:788–790.

66 Holloway JW, Doull I, Begishvili B, Beasley R, Holgate ST, Howell WM: Lack of evidence for a significant association between HLA-DR, -DQ and -DP genotypes and atopy in families with HDM allergy. Clin Exp Allergy 1996 (in press).

67 Standring P, Warner JA, Warner JO, Howell WM: DRB8-DQB4 association with non-atopic status in families with two or three generations of asthma. Eur J Immunogenet 1996;23:83.

68 Amelung PJ, Panhuysen CIM, Postma DS, Levitt RC, Koeter GH, Francomano CA, Bleecker ER, Meyers DA: Atopy and bronchial hyperresponsiveness: Exclusion of linkage to markers on chromosomes 11q and 6p. Clin Exp Allergy 1992;22:1077–1084.

69 Higgins JA, Thorpe CJ, Hayball JD, O'Hehir RE, Lamb JR: Overlapping T-cell epitopes in the group I allergen of *Dermatophagoides* species restricted by HLA-DP and HLA-DR class II molecules. J Allergy Clin Immunol 1994;93:891–899.

70 Reed EF, Tugulea SL, Suciu-Foca N: Influence of HLA class I and class II antigens on the peripheral T-cell receptor repertoire. Hum Immunol 1994;40:111–122.

Dr. W.M. Howell, Molecular Immunology Group, Tenovus Laboratory,
Southampton University Hospitals, Tremona Road, Southampton SO16 6YD (UK)

Hall, IP (ed): Genetics of Asthma and Atopy.
Monogr Allergy. Basel, Karger, 1996, vol 33, pp 71–96

....
The Genetics of Specific Allergy

Miriam F. Moffatt, William O.C.M. Cookson

Asthma Genetics Group, Nuffield Department of Clinical Medicine, John Radcliffe
Hospital, Oxford, UK

Introduction

Atopic asthma is precipitated by common inhaled proteins, known as
allergens. Although 50% of the general population may have positive skin
tests to common allergens [1–3], far fewer than this number are asthmatic
[3–5]. This is in part because atopic individuals differ in the allergens to which
they react. Asthma and bronchial hyperresponsiveness are associated primarily
with allergy to house dust mite (HDM), and in a lesser degree to cat dander
and to moulds such as *Alternaria* spp. [4, 6]. Grass pollen allergy, on the other
hand, carries no increased risk of asthma [4, 6].

The ability to react to a particular allergen is determined by genetic and
environmental factors. Understanding how these factors interact may help in
the prevention of allergy, or the induction of tolerance to particular allergens.
In addition, study of the genetic factors may give a general insight into the
response of the immune system to exogenous antigens.

Humoral Immunity

In a simple version of humoral immunity, immature B cells produce
immunoglobulin which is expressed on the cell surface, independently of any
effects from antigen. Although an individual B cell and its progeny are commit-
ted to making a specific immunoglobulin, immunoglobulins are constructed
with a random affinity for antigen. When a B cell contacts exogenous antigen
which binds to its immunoglobulin, the antigen is taken into the cell. Within
the cell the antigen is processed into small peptides, approximately 12–17

amino acids in length, which are carried back to the cell surface and presented with a molecule from the class II region of the major histocompatibility complex (MHC). The combination of the peptide and MHC is recognized by a helper T cell, through the T-cell receptor (TCR). Helper T cells typically carry a transmembrane protein known as CD4. Helper T cells control the subsequent growth and division of the B cell, through to the production of immunoglobulin-secreting plasma cells. T-cell control of B cells may be 'cognate', when T and B cell are in contact, or 'noncognate' through the secretion of cytokines. The T cell itself requires activation signals, usually by a reaction with dendritic cells. These cells, of which Langerhans' cells in the skin are examples, bind immunoglobulins to their cell surface by Fc receptors. Antigen bound to the immunoglobulin is then taken into the cell. After processing the antigen into peptides, dendritic cells present large numbers of MHC-peptide complexes on their cell membranes. Once activated by dendritic cells, T cells can react independently with B cells and other antigen presenting cells such as macrophages.

Thus, although T cells control B cells, the two cell types are recognizing antigens in quite different ways. Both B and T cells only react to parts of the foreign antigen, which are known as antigenic determinants, or 'epitopes'. The B cell, through its surface immunoglobulin, binds to an intact protein, whereas the T cell recognizes a short peptide in the context of a particular MHC molecule. B-cell epitopes are sensitive to the three-dimensional structure of antigen, so that they recognize 'conformational epitopes', whereas T cells recognize peptide sequences or 'sequential epitopes'.

Although immunoglobulins have an enormous range of possible affinities for different antigens, the interaction between B cell and T cell is more constrained. This is because the variation (polymorphism) in the MHC and TCR, and their ability to react to particular peptides or peptide complexes, is limited. This limitation is known as restriction. Restriction of the ability to respond to an antigen is the key to preventing allergy or to inducing tolerance for that antigen.

There are two classes of genes which may restrict specific IgE reactions. These are the genes encoding the human leucocyte antigen (HLA) proteins, from the MHC, and the genes for the TCR. The molecules encoded by these genes are central to the handling and recognition of exogenous antigen.

In this chapter, we will discuss the structure, function and genes of the MHC and the TCR, and review studies of their actions in specific allergy. First, however, we will consider the form and nature of allergens, and last we will examine the effects of environment.

Allergens and Epitopes

Inhaled allergen sources such as HDM are complex mixtures of many proteins. 'Major allergens' are proteins which consistently induce IgE responses in individuals reacting to an antigen source. Major allergens include *Der p* I (25 kD) and *Der p* II (14 kD) from the HDM *Dermatophagoides pteronyssinus*. *Alt a* I (28 kD) from the mould *Alternaria alternata, Can f* I (25 kD) from the dog *Canis familiaris, Fel d* I (18 kD) from the cat *Felis domesticus, Bet v* I (17 kD) from pollen of the birch tree (*Betula verrucosa*), and *Phl p* I (30 kD) and *Phl p* V (32 kD) from the pollen of timothy grass, *Phleum pratense*. Many other grass pollens, such as rye grass (*Lolium perenne*), have class I and class V allergens which are similar to those of *P. pratense*. It is not known why particular molecules act as allergens and induce IgE reactions, except that they are soluble, and that they find their way readily to the respiratory mucosa. It is likely that genetic associations will be better detected with reactions to purified major allergens, rather than with complex allergen sources.

B-cell epitopes have not been well defined for any of the major allergens. This is because antibody binding requires an intact protein, and it is difficult to dismember proteins into antigenic determinants without losing their three-dimensional structure. However, by using panels of mouse monoclonal IgE antibodies, *Der p* I has been shown to have four major B-cell epitopes [7], which were also recognized by human polyclonal IgE. Four murine IgG epitopes on *Der p* I were also found by other workers [8], but these did not cross-react with human IgE. *Der p* I epitopes are sensitive to denaturation, but *Der p* II antibody binding seems less dependent on protein conformation [9]. Murine IgG monoclonals have been used to show the presence of four B-cell epitopes on *Fel d* I, which were recognized by human polyclonal IgE and IgG [10]. Grass group I and V B-cell epitopes have not yet been mapped.

T-cell epitopes have been extensively studied for *Bet v* I [11, 12], and *Der p* I [13, 14] and *Der p* II [15, 16]. Although some favoured peptide segments for T-cell recognition have been reported near the N terminus of *Der p* I [13], other studies have found epitopes scattered along the molecule [14]. Similarly, epitopes seem widespread along *Bet v* I [11, 12] and *Der p* II [15, 16]. The presence of common epitopes for particular antigens, to which most individuals react, would be important for therapeutic strategies aimed at T-cell tolerance [16]. However, epitope mapping has not yet been carried out in a sufficient number of subjects to determine if such 'major epitopes' exist.

The Major Histocompatibility Complex

The genes encoding the HLA molecules are located within a 3,800-kb region, known as the MHC, on the short arm of chromosome 6 [17]. The HLA region is the most polymorphic gene system known in man. The polymorphism is due not only to the existence of multiple forms (alleles) at each locus, but also to the presence of several closely related genes at separate loci. There are three major gene clusters in the MHC, known as Class I, II and III. Classes I and II play a considerable role in T-cell immunity, by influencing the expressed T-cell repertoire [18], and by assuming a central function in antigen presentation [19].

The nomenclature of the MHC genes is quite complex, and invariably under revision as new products and allelic forms are constantly being identified. Currently, Class I genes are denoted by the suffix HLA-A to HLA-L, whilst Class II genes are suffixed HLA-DR, DQ, DP, DO, DN or DM to indicate their subregion of localization. Additionally, for Class II, an indication is given as to whether they encode α-chain sequences (i.e. HLA-DRA) or β-chain sequences (i.e. HLA-DRB). For allelic identification, the locus in question is followed by an asterisk and then the allele number and if required the subtype of the allele is also indicated (i.e. HLA-A*0201) [20].

Class I

HLA-A, B and C are termed 'classical' Class I genes, with the rest known as 'non-classical'. The classical genes exhibit substantial polymorphism [20]. The proteins encoded by the genes and expressed on the cell surface are known as 'antigens' (because they were originally identified by antibodies or mixed lymphocyte cultures). The antigens are characterized by a central groove or cleft, known as the antigen-binding site, in which digested peptide is presented to the TCR.

The HLA antigens consist of membrane-bound heterodimers (mixed twosomes) of two glycoprotein chains. The HLA genes encode a 45-kD heavy chain, which is non-covalently associated with a 12-kD serum protein known as beta-2-microglobulin (β2m). β2m is nonpolymorphic and encoded by a gene on chromosome 15. The heavy chain consists of three membrane-external domains (α1, α2 and α3), as well as a transmembrane portion and a cytoplasmic tail. The polymorphisms arising from the various allelic forms of the classical genes mainly localize within the α1 and α2 domains. Structure studies have shown that the polymorphisms are grouped together in and around the peptide-binding site, and that they are likely to be of functional significance.

The Class I classical molecules are expressed on nearly all nucleated cells [21]. In order for them to present antigen, it is necessary for the antigen to be synthesized within the presenting cells [19]. They play a central role in the recognition of antigen by T cells which carry the surface marker CD8. The majority of CD8+ T cells are cytotoxic, and are capable of lysing virally-infected cells.

The pattern of expression of the non-classical Class I genes has been greatly hampered by the lack of specific antibodies. Their functional importance is presently unclear.

Class II

The Class II genes are encoded by the HLA-D region. Like Class I, the expressed Class II antigens are membrane-bound heterodimers of glycoprotein chains. They consist of a 35-kD α chain and a 28-kD β chain that are noncovalently associated. Both chains have two membrane-external domains (α1 and α2, and β1 and β2 respectively), and each chain has a transmembrane portion and a cytoplasmic tail. The principal Class II molecules are HLA-DR, HLA-DQ and HLA-DP, each of which arises from α (e.g. DRA) and β (e.g. DRB) chain genes. The three principal Class II antigens differ in their degree of allelic polymorphism (see below).

Other Class II genes include HLA-DOB and HLA-DNA. No polypeptides have been detected for these genes, although their expression has been detected in mRNA [22, 23]. The HLA-DMA and DMB genes have been sequenced [24] although as yet they have no known function. Structural studies based on molecular modelling suggest that they may have a role in immune recognition.

Unlike Class I, the tissue distribution of the HLA Class II antigens is limited [25], but can also be augmented or induced by cytokines such as IFN-γ, TNF-α, and IL-4 [26]. The three main subclasses of Class II (DR, DP, DQ) are found on B cells, monocytes, macrophages, dendritic cells and activated T cells. HLA-DR, DQ and DP cell surface expression is also variable with a tendency for HLA-DR to be found at the highest levels, and HLA-DP at the lowest, although this is not always the case [37].

The Class II molecules present antigen that has been internalized by phagocytosis or membrane-bound molecules and processed within the cell [19]. Presentation of peptides is primarily to CD4+ T cells which are mostly of the 'helper' phenotype. The antigen presenting groove (cleft) for each Class II antigen will only accommodate peptides of a particular conformation. Peptides are anchored at their N and C termini into pockets within the MHC antigen cleft. As with Class I antigens, the majority of the variation coded for by all the Class II loci is clustered into three regions in and around the antigen-binding cleft [28]. Polymorphism in the cleft and pockets results in a differential ability to bind particular peptides, and is the structural basis of HLA restriction of the response to foreign antigens.

HLA-DR

The HLA-DR region contains a single DRA gene but multiple DRB genes. The DRA gene shows limited polymorphism, with only two recognized alleles [29, 30]. In contrast, the DRB genes are highly polymorphic. In addition to allelic variation within the DRB genes, more than one DRB gene may be expressed, and the number of DRB loci within the subregion can also vary. At present, a total of nine DRB loci have been characterized, of which five are pseudogenes. The number and type of DRB loci on individual chromosomes falls into distinct combinations, known as haplotypes. Most DR haplotypes express two DRB genes, with only two haplotypes expressing a single gene. The DRB1 gene is present on all haplotypes, and is the most polymorphic, with 60 alleles currently described. The second DRB locus (DRB3, DRB4 or DRB5) varies between haplotypes. All three secondary loci exhibit polymorphism, which is more limited than that seen for DRB1 [20].

HLA-DQ

The HLA-DQ region contains four genes grouped into two pairs. DQA1 and DQB1 genes are both expressed, giving rise to a glycoprotein heterodimer. Although mRNA transcripts may be detected from the other pair of genes, DQA2 and DQB2, it seems that they do not produce protein products. Both the DQA1 and DQB1 genes are polymorphic, with 15 DQA alleles and 26 DQB alleles currently recognized [20]. Unlike HLA-DR, the DQ polymorphisms do not all localize within the antigen-binding cleft, although most of the polymorphism in DQB is within the $\beta 1$ domain [28].

HLA-DP

The HLA-DP region also contains four genes grouped into two pairs. DPA2 and DPB2 are nonfunctional pseudogenes [31] whilst DPA1 and DPB1 result in the expressed heterodimer. A high number of alleles (59 to date) are found at the DNA level for DPB1, but the resulting amino acid variation is less than that seen for the DRB and DQB genes. DPA1 variability is limited, with eight alleles known [20].

Other Genes within the MHC

The MHC contains many genes associated with immune function, including complement genes and genes which code for the transportation of exogenous peptide within the cell [17]. These transporter peptides, known as TAP1 and TAP2, are found within the class II region, although their main influence seems to be on class I function [32]. Polymorphisms have been found in these genes, but their significance has not been tested in allergic disease.

Linkage Disequilibrium and the MHC

When a new polymorphism arises in a gene, it will be physically associated or linked with variants (alleles) of other genes and sequences on the same chromosome. This combination of alleles on a chromosome forms a haplotype. As the gene is transmitted through subsequent generations, the meiotic process of crossover and recombination means that alleles in the surrounding haplotype will gradually become unlinked. After an infinite number of meioses, the polymorphism in the gene will be randomly associated with all other alleles on the chromosome (i.e., in linkage equilibrium). However, when the number of meioses is not infinite, the polymorphism will still be linked to alleles which are closest to it on the chromosome (as crossover and recombination is less likely to have occurred in a short segment of DNA). The polymorphism is then in 'linkage disequilibrium' with adjacent alleles.

Linkage disequilibrium is seen in the MHC, and particular combinations of alleles or haplotypes are seen more often than would be expected by chance [33]. The strength of disequilibrium suggests that there is an evolutionary advantage to conserving particular haplotypes, although the reasons for this are not known.

HLA-DR and HLA-DQ are in strong linkage disequilibrium. DQ lies between DR and DP and all three are contained within approximately 1 Mb [34]. However, only weak linkage disequilibrium exists between DR and DP. The two genes of DQ also differ in their strength of linkage disequilibrium to DRB, with DQA having stronger associations than DQB. Although linkage disequilibrium is not unique to the HLA-D region, it is most notable for this particular region of the MHC. The presence of linkage disequilibrium means that HLA-DR and DQ associations may be difficult to tell apart, but the studies described below suggest that HLA-DR has the most important effects on specific allergy.

HLA and Allergy

There have been numerous studies of HLA and IgE responses to different allergens (see previous chapter). In general, the studies have often been of small numbers of subjects, and positive results have not been confirmed in other individuals. A substantial problem with case-control HLA association studies is that the many different HLA alleles lead easily to multiple tests of significance. If, for example, 16 common DRB alleles are recognized by a particular technique, then an association study is immediately testing 16 independent hypotheses, and the p value from any test should be corrected by multiplying by 16. Any subdivision of the data that was not intended a priori also influences the level of significance. In the above example, dividing the subjects into two groups means that p should be multiplied by 32. Unfortunately, these stringent criteria are infrequently applied.

The evolving nomenclature of the HLA genes and antigens potentially confuses a review of HLA association studies. In the following paragraphs we give the HLA type at the time of the original reports, and if this is ambiguous, we also give the modern classification.

HLA restriction of the IgE response was first noted by Levine et al. [35], who in a study of families found an association between HLA haplotypes and IgE responses to antigen E derived from ragweed (*Ambrosia artemisifolia*). This finding was subsequently shown to be due to the association of the minor ragweed allergen *Amb a* V (molecular weight 5,000 kD) and HLA-DR2 [36]. This association has been consistently confirmed [37, 38], and has been refined to the DR2 subtype HLA-DRB1*15 [38]. An HLA-DR5 association with another ragweed antigen *Amb a* VI has been reported in 38 subjects, but not confirmed [39].

An association between *Alt a* I and HLA-DRB1*04 has been reported in a study of 430 British subjects in 84 families recruited through allergy clinics [40]. Genetic linkage of IgE responses to the allergen and the HLA region was also noted in these subjects [40]. In a multicentre international study, this same association was seen in Bulgarians, Israelis and Japanese [38], albeit at a marginal level of significance in each case.

A positive association between *Bet v* I, the major allergen of birch pollen, and DRw52 (HLA-DRB3*0101) has been reported in 37 Austrian subjects [41], and confirmed in 41 Danish subjects [42].

Other suggested associations are of the rye grass antigens *Lol p* I, *Lol p* II and *Lol p* III with HLA-DR3 (in the same 53 allergic subjects) [43, 44], IgE reactions to American feverfew (*Parthenium hysterophorus*) and HLA-DR3 in 22 subjects from the Indian subcontinent [45], HLA-DRB1*01 and *Fel d* I in 430 British subjects [40], and DR7 and DQ2 and *Ole e* I (from the olive tree) in 40 Spanish subjects [46].

Other authors have reported negative associations with particular allergens. These include HLA-DR4 and IgE responses to mountain cedar pollen (37 subjects) [47] and HLA-DR4 and melittin (from bee venom) (22 subjects) [48]. Nonresponsiveness to Japanese cedar pollen may be associated with HLA-DQw8 [49].

Thus three associations seem secure: *Amb a* V and DRB1*15 (DR2), *Alt a* I and DRB1*04, and *Bet v* I and DRB3*0101. There is to date no confirmation of other possible positive findings, which in general have been carried out on too few subjects to establish an unequivocal HLA assocation. In addition, the large environmental differences between populations have not been studied systematically. It is of note that no consistent associations have been found for either *Der p* I or *Der p* II, even though HDM is ubiquitous in most westernized societies. In addition, there has not been recognition of the problems of reactivity to multiple allergens: significant relationships between HLA-DR alleles and five antigens (*Amb a* V, *Lol p* I, *Lol p* II, *Lol p* III and *Amb a* VI) have been claimed from the same pool of approximately 200 subjects [36, 39, 43, 44]. These difficulties may be dealt with in the future by regression analysis [38, 40].

Apart from *Amb a* V and DRB1*15, the relative risk for positive associations is generally around 2 [40], and is well below that generally seen with HLA associations and autoimmune diseases. In vitro studies of T-cell reactions to antigen presenting cells have shown multiple Class II molecules may present the same antigen-derived peptide [50, 51], and therefore it seems likely that the HLA system is not the major limiter of specific IgE reactions.

The T Cell and Its Receptor

Given that the MHC cannot fully account for restriction of IgE responses, the highly variable distal domains of the TCR, which recognize the complex of MHC and peptide on antigen presenting cells, become important candidates for modulating the response to particular allergens.

Structure

The TCR consists of a disulphide-linked heterodimer of two highly polymorphic glycoprotein chains, known as α and β. The receptor is noncovalently associated with three other molecules that constitute the CD3 complex [52] as well as two signal transduction elements, η and ζ. CD3 is involved in TCR assembly and cell surface expression [53, 54] and also plays a role in signal transduction following TCR engagement [55, 56].

The class of MHC with which the TCR interacts depends upon the expression of its co-receptor, which may be either CD4 or CD8. The co-receptor is essential for T-cell development [57, 58]. Commitment of a T cell to express CD4 results in it being Class II restricted, whilst expression of CD8 results in Class I restriction. The majority of CD4+ T cells are of the helper phenotype whilst CD8+ cells tend to be of the cytotoxic phenotype [59].

As well as a role in thymic differentiation, CD4 and CD8 may also have a part to play during primary encounter of antigen by a naïve T cell, since both in vitro and in vivo studies have shown disruption of normal T-cell priming when the co-receptors are absent or blocked [60, 61]. Once T cells are primed, engagement of CD4 and CD8 may not be a requirement for T-cell activation, although they enable recognition to occur efficiently with lower concentrations of antigen [62, 63].

The co-receptors may also play a role in TCR signalling when they are stimulated [64–67]. Both molecules are associated with a tyrosine kinase that may have an involvement in signalling events [68, 69].

A second type of TCR, consisting of γ and δ chains, was discovered whilst obtaining primary sequences of the α and β chains from cDNA and genomic clones [70, 71]. Like the αβ heterodimer, the γδ TCR is also expressed in association with the CD3 complex [72]. In humans, the γδ T-cell population only accounts for 1–10% of peripheral T cells [73–75].

The function of the γδ T cells has not been fully resolved. Unlike αβ T cells, they do not apear to recognize processed antigen associated with Class I and Class II MHC molecules. However, it has been suggested that populations of the cells may have a specific role in bacterial and parasitic infection [76–79], or recognize generically stressed cells [80, 81]. In the neonate, the cells may be required before complete development of the TCRαβ repertoire occurs [82, 83].

γδ T cells may be of importance in controlling IgE regulation, since populations have been found on mucosal surfaces, such as the epithelium of the lung, where allergens initiate IgE responses [84, 85]. It has been suggested that γδ T cells may protect the respiratory epithelium from inflammatory damage by selectively suppressing IgE responses [84]. Experiments involving ovalbumin-sensitized mice have shown that antigen-specific γδ T cells selectively suppress the Th2-dependent IgE antibody production without affecting parallel IgG responses [86].

Genomic Organization and Diversity of T-Cell Repertoire

The α-chain locus is on the long arm of chromosome 14 [87, 88] and contains within it the δ-chain locus [71, 89]. Both the β-chain locus and the γ-chain locus are on chromosome 7 [90, 91]. The genomic organization of the TCR genes is similar to that of immunoglobulin [92]. Each locus is divided into an array of interchangeable coding segments scattered over large tracts of chromosomal DNA. Before the genes become functional, the variable (V), diversity (D), joining (J) and constant (C) segments are separate in the genomic (germline) DNA. During thymic maturation of a T cell, excision of intervening DNA and recombination brings together the segments to produce a contiguous VJ exon in the case of the α and γ chains, and a contiguous VDJ exon for β and δ chains. RNA processing results in splicing of a C segment to the exon and translation produces the glycoprotein chains. Like immunoglobulin, the TCR contains three or four hypervariable loops that correspond to the complementarity-determining regions (CDR) [93]. The CDR1 and CDR2 loops of the TCR are encoded by the V gene region with the CDR3 encoded by the V-(D)-J junctional regions [92, 93].

The large number of gene segments for the TCR loci means that combinatorial diversity can generate many transcriptional permutations, which are the primary source of receptor diversity. The degree of potential variability is increased by the random associations of different α and β chains, or γ and δ. Junctional variation also occurs, with the addition of nongermline encoded nucleotides at the N regions by terminal deoxytransferase. This enzyme is highly active in the thymus, the site of T-cell maturation.

It is estimated that generation of 10^{16} different receptors by this system is possible, and that this is sufficient to account for the vast array of antigens that the immune system might encounter [92].

However, studies of peripheral blood lymphocytes in unrelated individuals have shown that the expressed TCR repertoire (the usage of particular V, J or D segments) is not random. Individual variation has been seen in the use of the V and J segments of the TCR-β locus, and the V segments of the TCR-α locus [94, 95]. The repertoire is known to be influenced by genetic factors, particularly the MHC [96–98]. It has therefore been suggested that differences in genomic organization may shape the peripheral T-cell repertoire [99, 100].

This possibility of germline variation (genomic polymorphism) influencing an individual's response to antigens, has resulted in studies of association between TCR gene polymorphisms and human disease [101–103].

A number of allelic polymorphisms within the TCR loci have been identified, the majority being within the V gene segments. Restriction fragment length polymorphisms (RFLP) mapping to noncoding regions have also been characterized. These have played a key role in disease association studies, with haplotypes being constructed after RFLP typing [104–106]. Structural V gene polymorphisms have also been identified, the first being in Vβ6.7 gene segment [107]. A modest number of coding Vβ polymorphisms are now known [101, 108–112]. On the basis of cDNA sequencing, it has been suggested that a number of other Vβ gene segments may be polymorphic [94, 109].

The first example of a frequent coding polymorphism for the TCR-α locus was for the Vα8.1 segment [110]. Subsequent investigation by RFLP, SSCP (single-stranded conformational polymorphism) and DGGE (denaturing gradient gel electrophoresis) techniques has resulted in further coding and noncoding polymorphisms being identified in a range of Vα segments [113–117]. Little is known about allelic polymorphism in the TCR-γ and δ genes.

A further dimension to the variability of the TCR genes complexes is offered by insertion or deletion polymorphisms which have been found in the

TCR-β gene complex, and involve stretches of 15–30 kb of DNA [118]. By contrast, there would appear to be few major insertions or deletions within the TCR-α locus.

Any allelic variation that causes an amino acid substitution may alter the receptor's ability to recognize antigen. In the case of immunoglobulin, amino acid changes have been found to affect the conformation of the hypervariable regions of the molecule and to increase antigen binding [119]. Marked differences in α and β chain pairing due to allelic variation have been observed, with the V region of the β chain being the determining factor for pairing ability [120]. Junctional diversity in the VDJ junction of the TCR has been shown to alter the specificity of antigen recognition by the receptor [121, 122].

Although genetic factors predominate, nongenetic factors have a detectable effect on the selection of the TCR repertoire. Twin studies, in which discordance for particular Vβ families has been seen in monozygotic twins despite their identical genetic backgrounds, suggest that the environment also influences the TCR repertoire [94, 123, 124].

Restriction of the Expressed TCRA and TCRB Gene Repertoire and Its Role in Ig Production

Restriction of V gene usage and conservation of junctional sequences or both, have been shown by a number of studies in which TCRs recognize defined peptide/MHC complexes. Experimental allergic encephalomyelitis (EAE) is a disease in inbred mice which shares some features of human multiple sclerosis (MS). Studies of murine CD4+ T-cell clones which mediated EAE found that all the analyzed clones used the same Vα4 gene segment, with at least 80% using the same Vβ8 gene segment. Most of the clones used the same Jα gene segment, but the use of Jβ gene segments was more heterogeneous. However, the clones could still be classed into four distinct groups based on their Vβ-Jβ combinations [125]. These results illustrate a strong in vivo selection for particular TCR elements.

Immunization with a synthetic peptide derived from a hypervariable region of the TCR Vβ8 molecule was found to confer passive protection against EAE in naïve animals. It was shown that immunization with the peptide had resulted in the proliferation of Vβ8-specific regulatory T cells that prevented the induction of the disease [126].

However, in human MS, studies of T cells specific for myelin basic protein (MBP) did not find selective usage of TCR V region genes. However, human MBP-specific T-cell clones may recognize a much broader spectrum of MBP epitopes than their rodent counterparts. It is therefore likely to be much more difficult to establish if a restricted T-cell repertoire exists in patients with MS [127, 128].

Extensive sequence comparison of α and β chains expressed by murine T cells specific for pigeon cytochrome c showed restricted V gene usage for both chains. A striking selection for junctional sequences in β chains was seen, but there appeared to be no such selection in the α chains [129]. A number of other studies have also reported that murine T cells which recognize particular peptide/MHC complexes similarly share particular V region or J region sequences [121, 130, 131].

In humans, investigations of TCRs specific for the HLA-A2 and influenza A matrix peptide complex have found considerable conservation of Vα and Vβ sequences, as well as some junctional sequence conservation. When another influenza peptide (influenza B nucleoprotein 82–94), which is also HLA-A2 restricted, was examined, none of the sequences of the influenza A responses were seen. Repertoire studies showed the conserved Vα and Vβ sequences to be infrequent in the peripheral blood, so that strong selection had occurred for them to be so prevalent in the influenza A-specific clones [132].

Using a bacterial entertoxin-based proliferation assay and cDNA sequencing to identify V and J sequences. Boitel et al. [133] found preferential usage of a particular Vβ region gene segment in human TCRs specific for a tetanus toxin-derived peptide. However, clones having identical Vβ genes did not share the same MHC Class II restriction. No consistent pattern of Vα gene usage was observed and there was also no apparent conservation of junctional sequence even for TCRs with identical Vβ gene segments. A later study of another bacterial derived peptide, mycobacterial heat-shock protein, also found a predominant usage of particular Vα and Vβ gene segments and additionally, the T-cell clones were found to share the same MHC Class II restriction [134].

In human autoimmune disease, isolation of T cells directly from disease sites has also shown oligoclonality of the V gene usage although the antigen has been unknown. Synovial fluid taken from 2 patients suffering from rheumatoid arthritis was analysed for the relative expression of TCR Vα and Vβ segments using PCR-based techniques. It was found that T cells at the site of inflammation expressed a limited number of Vβ gene segments compared to those expressed by peripheral blood lymphocytes. The repertoire of Vα gene segments was remarkably similar for both peripheral lymphocytes and synovial fluid lymphocytes [135].

In autoimmune thyroid disease, intrathyroidal T cells as well as peripheral blood T cells were examined for their Vα usage. Eighteen known Vα gene families were examined by polymerase chain reaction in 9 patients. Whilst peripheral blood T cells expressed 17 of the Vα genes, in each patient an average of only 5 of the 18 Vα genes were expressed by T cells from the thyroid [136].

However, for both rheumatoid arthritis and autoimmune thyroid disease, a number of further studies have reported a variety of Vα and Vβ segments to have been expanded or deleted in patients suffering from the illnesses [137–141]. The discrepancies in the results, and the lack of reproducibility makes it difficult to decide if the TCR repertoire is restricted in these diseases.

T-cell clones have been studied for responses to a number of respiratory allergens. The first of these was by Wedderburn et al. [142] who examined the TCR sequences of a panel of HDM-specific T-cell clones from an individual with perennial rhinitis. Ten T-cell clones were isolated and their antigen specificity determined. Six were specific to *Dermatophagoides farinae,* the inducing antigen, whilst the other four showed cross-reactivity to the closely related HDM species, *D. pteronysinnus.* The V gene usage of the clones was examined by cloning and sequencing of amplified TCR gene products. Of the ten clones, six had unique TCR sequences, with the remaining four being identical to others in the panel. A restricted number of TCR V genes were used, with a dominant expression of Vβ3 and Vα8 genes. There was considerable heterogeneity of the J gene segments. By analysing the TCR segments of clones isolated in different years, the investigators concluded that long-lived dominant T-cell clones exist in vivo, presumably maintained by chronic allergen exposure.

A restricted V gene usage was similarly observed in a study of T-cell clones cross-reactive for the major grass pollen allergens *Lol p* I of rye grass (*L. perenne*) and *Poa p* IX of Kentucky bluegrass (*Poa pratensis*) [143]. Using RNA-PCR followed by cloning and sequencing of the amplified Vα gene products, nine out of ten *Lol p* I-specific T-cell clones were found to possess the Vα13 gene. All contained different Jα gene sequences. Vβ gene usage was only examined for two clones. Interestingly, in analysis of seven *Poa p* IX-specific T-cell clones, five were again found to utilize the Vα13 gene segment. It is therefore possible that a limited number of TCR Vα genes are used by grass-allergen-specific T-cell clones.

The diversity of Vα and Vβ sequences used by T-cell clones specific for the major birch tree pollen allergen, *Bet v* I, has also be examined by PCR-based techniques. Five CD4+ T-cell clones, each from a separate donor, and each specific for the major epitope of *Bet v* I, were studied. Although all the donors possessed the DRB1*07 genotype, T-cell reactivity with *Bet v* I was only found to be restricted by this HLA molecule in one clone. Only two of the five clones used an identical Vβ gene segment with each of the remaining three clones expressing a different Vβ family member. Two of the clones expressed closely related Vα gene segments but other Vα gene segments were drawn from different families. A diverse range of junctional gene segments were used in α and β chains, with no two clones using the same segments [144].

A strong association between HLA-DR2 and Dw2 (HLA-DRB1*15) and specific IgE responses to *Amb a* V has been shown previously [36, 38]. Subsequent studies of *Amb a* V-specific T-cell clones have confirmed that proliferative responses were only seen when HLA-DRB1*15 was present on the antigen presenting cells [145]. Further studies of the TCRs of the T-cell clones found Vβ5.2 to be dominant in polyclonal *Amb a* V-specific cell lines, although it occurred at a low frequency in peripheral blood lymphocytes [146]. Restriction of Vβ usage in *Amb a* V-specific T-cell clones is supported by earlier studies in a mouse model of T- and B-cell responses after sensitization to a ragweed extract, in which specific Vβ T-cell subsets mediated the immediate hypersensitivity response [147].

From a clonal level, it would therefore appear that there is restricted usage of TCR Vα and Vβ genes for a number of major respiratory allergens. The effects of restriction on IgE production has as yet to be fully assessed in humans, but it may be that different V gene usage is associated with specific patterns of cytokine production in T-cell subsets [148]. Animal experiments suggest that T cells bearing different Vβ elements are differentially involved in the regulation of Ig production.

Studies in the chicken, for which there are only two Vβ gene families (1 and 2), showed that IgA production was selectively and severely compromised in Vβ1-depleted birds. Production of IgM and IgG antibodies was not impaired. This implied that B-cell production of IgA had a requirement for T cells expressing Vβ1 genes [149].

TCR Vβ elements have also been shown to regulate IgE production in a murine model of ovalbumin-induced airways hyperresponsiveness (AR). In this model, expansion of Vβ8.1/8.2 T cells was found in the draining lymph nodes of the airways and the lung. In vitro these cells augmented IgE production, whilst Vβ2 cells limited the increase. Transfer experiments confirmed the in vitro observations, with naïve mice gaining the capacity to develop allergen-specific IgE, immediate cutaneous hypersensitivity and altered AR on receipt of Vβ8.1/8.2 cells. Co-transfer of Vβ2 cells prevented the allergic response [150, 151].

In mice sensitized to ragweed allergen, co-culture of specific T cells with ragweed primed B cells produced different stimulation profiles for Ig isotypes which depended upon which Vβ element was expressed. Polyclonal Vβ8.2 T cells stimulated IgE and IgG1 production and this was confirmed by transfer experiments of the cells from sensitized mice to naïve recipients. Tranfer also increased airways responsiveness in the recipients. Cells carrying other Vβ elements did not enhance IgE production, and no such effects were seen when Vβ8.2 cells were transferred from nonsensitized control mice [147].

The above experiments suggest that restricted receptor repertoires do exist for T cells specific for respiratory allergens, and repertoire usage directly affects the production of IgE. However, many of the studies are based on only a small number of T-cell clones from a limited number of individuals, which makes it difficult to fully interpret the results. In addition there are technical difficulties in carrying out repertoire studies.

Analysis of the TCR repertoire can best be carried out by using monoclonal antibodies against particular V or J segments. However, clonotypic antibodies for T cells have been difficult to obtain, with very few characterized for Vα and only 50% of the expressed Vβ repertoire covered. The use of antibodies is advantageous because they give reproducible results, but they do not yield sequence information. An alternative approach, utilized by many of the studies detailed above, is based on PCR [152, 153], particularly of specific V region families [154, 155]. V region family-specific PCR is the quickest technique, but variation in the efficiency of the PCR, depending on the sets of primers used, may lead to artefact in the results. In general, PCR-based techniques have been less reproducible than antibody analysis, and studies will become more reliable as more antibodies against a wider range of Vα and β segments become available.

Genetic Linkage

An alternative approach to determining if the TCR genes influence susceptibility to particular allergens, is to test whether genetic linkage exists between an allergic response and the chromosomal loci encoding the α and β chains of the TCR. When present, genetic linkage broadly localizes a genetic effect to a chromosomal region, without defining the exact genes or sequences causing the effect. Once genetic linkage has been established, individual allelic polymorphisms in genes can then be examined for associations with specific IgE responses. Genetic linkage is best studied with highly polymorphic markers. Microsatellite markers associated with the T-cell loci have been identified [107, 156].

Genetic linkage has been shown between the TCR-α/δ locus and specific IgE responses to major allergens, but has not been found for the TCR-β locus [157]. Two independent sets of families were examined, one from the United Kingdom and the other from the western coastal town of Busselton in Australia. In total, 823 subjects containing 312 sibling pairs were studied. All subjects were phenotyped for IgE titres to whole allergens as well as to a number of highly purified major allergens. Genotypes were determined for the two microsatellite DNA sequences close to the TCR loci, one on chromosome 14 the other on chromosome 7. Affected sib pair analysis showed no genetic linkage to the TCR-β locus. However, strong genetic linkage between the TCR-α/β locus and specific IgE responses was found. Linkage to the total serum IgE was also seen. Due to the close correlation between total and specific IgE, it is difficult to say whether the locus controls specific IgE reactions or confers general IgE responsiveness. However, the linkage was strongest with highly purified allergens (*Der p* I, *Der p* II and *Fel d* I) suggesting that the locus primarily influences specific IgE. There were no allelic associations with any of the microsatellite alleles. The linkage was significantly stronger in maternally derived alleles [158].

Genetic linkage of the total IgE to the TCR-α/β locus was also found in a set of 53 nuclear families taken from a random population sample of the urban and rural areas of the southern Black Forest in Germany [159]. This linkage was only seen for maternal alleles. No linkage for specific IgE responses was found, although testing with major allergens was not performed.

Moffatt et al. [160] continued their genetic linkage study by typing their subject sets for a bi-allelic polymorphism in the Vα8.1 gene. The subjects were also HLA-DRB typed so that interactions between HLA and TCR loci could be investigated. A strong allelic association with Vα8.1 and reactivity to *Der p* II was seen in the Australian subjects, but not in the British. The lack of association in the British set of families may be attributable to population differences in environmental exposure to HDM. Certainly, the Australian subjects had been exposed to a higher concentration of HDM in the mattress and had higher IgE responses to whole HDM as well as to the purified allergen *Der p* II. In order to investigate whether interacting effects with HLA-DRB existed for the association, multiple regression analysis with IgE titre to *Der p* II as the dependent variable and Vα8.1 and HLA-DRB types as independent variables was performed. The results showed that HLA-DRB1*02 was also positively associated with IgE titres to *Der p* II, consistent with a general hypothesis of interacting HLA-DR and TCR-α restriction of IgE response to particular antigens. Vα8 has been found to be dominant in T-cell clones reacting with HDM [142], but in order to determine whether the polymorphism in Vα8.1 is directly responsible for the allelic associations, further genetic as well as functional studies are required.

Environment

Even in genetically similar people, the dose and timing of allergen exposure will have important effects on subsequent manifestations of the atopy phenotype. Schoolchildren raised at high altitude, where there is little pollen, and where low humidity prevents HDM growth, develop less allergy than those raised at sea level [161]. Similarly, children living in the dry interior of Australia develop less allergy than those living in more humid conditions near the coast [3].

Events in early infancy are critical in determining the subsequent course of allergic disease. In the Scandinavian countries, a short intense spring flowering of birch trees is accompanied by symptoms in many individuals. Children born in the 3 months around the pollen season carry an increased risk of allergy to birch pollen for the rest of their life [162]. In English children the level of HDM in infants' bedding during the first year of life correlates with the subsequent risk of childhood asthma [163].

The increase in the prevalence of atopic disease in this century must be due to an environmental factor or factors. Air pollution has declined steadily since the 1950s in England and Western Europe, and ozone levels remain stable despite an increase in the number of cars. Comparative studies of the prevalence of asthma and atopy have been carried out between East and West Germany, two regions with genetically similar populations, and with far higher levels of atmospheric pollution in the East [164]. Surprisingly, the prevalence of asthma and skin test responses to allergens is lower in the East than in the West [165]. Similar results are seen when the Baltic States are compared to Sweden [166]. This difference may be attributable to childhood respiratory infections, as pollution and overcrowding predispose to infantile infection. Support for this hypothesis is given by the finding that the youngest children in large sibships have significantly less asthma than their older siblings [167]. At the cellular level, it is suggested that early infections program the immature immune system towards a Th1 rather than a Th2 helper cell profile, thereafter favouring cellular rather than IgE responses. Parasitism may also 'protect' against allergic disease [168, 169], perhaps by the production of polyclonal IgE which overwhelms specific IgE in competition for binding on effector cells.

Conclusions

Data from many different types of study show that polymorphism in the genes of the MHC and of the T-cell receptor influence the ability of the individual to respond to particular allergens. In contrast to autoimmune diseases, HLA effects in allergy are in general weak, but it is possible that combinations of MHC and TCR polymorphisms more strongly restrict the response to allergens. Immune tolerance to specific antigens is a feature of IgE-mediated allergy, and the induction of tolerance is an important therapeutic goal for the treatment of asthma and hay fever. Understanding the mechanisms underlying specific allergy, through genetic and functional studies, may be the key to rational prevention and treatment of the atopic diseases.

References

1   Cline MG, Burrows BB: Distribution of allergy in a population sample residing in Tuscon, Arizona. Thorax 1989;44:425–432.
2   Holford-Strevens V, Warren P, Wong C, Manfreda J: Serum total immunoglobulin E levels in Canadian adults. J Allergy Clin Immunol 1984;73:516–522.
3   Peat JK, Britton WJ, Salome CM, Woolcock AJ: Bronchial hyperresponsiveness in two populations of Australian school children. III. Effect of exposure to environmental allergens. Clin Allergy 1987; 17:271–281.
4   Sears MR, Herbison GP, Holdaway MD, Hewitt CJ, Flannery EM, Silva PA: The relative risks of sensitivity to grass pollen, house dust mite and cat dander in the development of childhood asthma. Clin Allergy 1989;18:419–424.
5   Peat JK, Haby M, Spijker J, Berry G, Woolcock AJ: Prevalence of asthma in adults in Busselton, Western Australia. Br Med J 1992;305:1326–1329.
6   Cookson WOCM, De Klerk NH, Ryan GR, James AL, Musk AW: Relative risks of bronchial hyper-responsiveness associated with skin-prick test responses to common antigens in young adults. Clin Exp Allergy 1991;21:473–479.
7   Lind P, Hansen OC, Horn N: The binding of mouse hybridoma and human IgE antibodies to the major faecal allergen, Der p I, of Dermatophagoides pteronyssinus. J Immunol 1988;140:4256–4262.
8   Chapman MD, Heymann PW, Platts-Mills TA: Epitope mapping of two major inhalant allergens, Der p I and Der f I, from mites of the genus Dermatophagoides. J Immunol 1987;139:1479–1484.
9   Lombardero M, Heymann PW, Platts-Mills TA, Fox JW, Chapman MD: Conformational stability of B cell epitopes on group I and group II Dermatophagoides spp. allergens. Effect of thermal and chemical denaturation on the binding of murine IgG and human IgE antibodies. J Immunol 1990; 144:1353–1360.
10 Vailes LD, Li Y, Bao Y, DeGroot H, Aalberse RC, Chapman MD: Fine specificity of B-cell epitopes on Felis domesticus allergen I (Fel d I): Effect of reduction and alkylation or deglycosylation on Fel d I structure and antibody binding. J Allergy Clin Immunol 1994;93:22–33.
11 Ebner C, Schenk S, Szépfalusi Z, Hoffmann K, Ferreira F, Wilheim M, Schneir O, Kraft D: Multiple T cell specificities for Bet v I, major birch pollen allergen, within single individuals. Studies using specific T cell clones and overlapping peptides Eur J Immunol 1993;23:1523–1527.
12 Ebner C, Szépfalusi Z, Ferreira F, Jilek A, Valenta R, Paronchi P, Maggi E, Romagnani S, Scheiner O, Kraft D: Identification of multiple T cell epitopes on Bet v I, the major birch pollen allergen, using specific T cell clones and overlapping peptides. J Immunol 1993;150:1047–1054.
13 O'Brien MR, Thomas RW, Tait DB: An immunogenetic analysis of T-cell reactive regions on the major allergen from the house dust mite, Der p I, with recombinant truncated fragments. J Allergy Clin Immunol 1994;93:628–634.
14 Yssel H, Johnson KE, Schneider PV, Wideman J, Terr A, Kastelein R, De Vries JE: T cell activation inducing epitopes of the house dust mite allergen Der p I. Proliferation and lymphokine production patterns by Der p I-specific CD4+ T cell clones. J Immunol 1992;148:738–745.
15 Van Neerven RJ, van t'Hof W, Ringrose JH, Jansen HM, Aalberse RC, Wierenga EA, Kapsenberg ML. T cell epitopes of house dust mite major allergen Der p II. J Immunol 1993;151:1–10.
16 O'Hehir RE, Verhoef A, Panagiotopoulou E, Keswani S, Hayball JD, Thomas WR, Lamb JR: Analysis of human T cell responses to the group II allergen of Dermatophagoides species: Localisation of major antigenic sites. J Allergy Clin Immunol 1993;92:105–113.
17 Trowsdale J, Ragoussis J, Campbell RD: Map of the human MHC. Immunol Today 1991;12: 443–446.
18 Schwartz RH: Acquisition of immunologic self-tolerance. Cell 1989;57:1073–1081.
19 Germain RN: The ins and outs of antigen processing and presentation. Nature 1986;322:687–689.
20 Bodmer JG, Marsh SGE, Albert ED, Bodmer WF, Dupont B, Erlich HA, Mach B, Mayr WR, Parham P, Sasazuki T, Schreuder GmTh, Strominger JL, Svejgaard A, Terasaki PI: Nomenclature for factors of the HLA system 1994. Human Immunol 1994;41:1–20.
21 Daar AS, Fuggle SV, Fabre JW, Ting A, Morris PJ: The detailed distribution of HLA-A, B, C antigens in normal human organs. Transplantation 1984;38:287–292.

22   Tonnelle C, DeMars R, Long EO: DOβ: A new β chain gene in HLA-D with a distinct regulation of expression. EMBO J 1985;4:2839–2847.

23   Trowsdale J, Kelly A: The human HLA class II α chain gene DZα is distinct from genes in the DP, DQ and DR subregions. EMBO J 1985;4:2231–2237.

24   Kelly AP, Monaco JJ, Cho S, Trowsdale J: A new human HLA-class II related locus, DM. Nature 1991;353:571–573.

25   Daar AS, Fuggle SV, Fabre JW, Ting A, Morris PJ: The detailed distribution of MHC class II antigens in normal human organs. Transplantation 1984;38:293–298.

26   Arenzana-Seisdedos F, Mogensen SC, Vuillier F, Fiers W, Virelizier J-L: Autocrine secretion of tumor necrosis factor under the influence of interferon-γ amplifies HLA-DR gene induction in human monocytes. Proc Natl Acad Sci USA 1988;85:6087–6091.

27   Lee JS: Regulation of HLA class II gene expression; in Dupont B (ed): Immunobiology of HLA. New York, Springer, 1989, vol II, pp 49–61.

28   Bell JI, Todd JA, McDevitt HO: Molecular structure of human class II antigens; in Dupont B (ed): Immunobiology of HLA. New York, Springer, 1989, vol IIB, pp 40–48.

29   Lee JS, Trowsdale J, Travers PJ, Carey J, Grosveld F, Jenkins J, Bodmer WF: Sequence of an HLA-DR α-chain cDNA clone and intron-exon organization of the corresponding gene. Nature 1982; 299:750–752.

30   Das HK, Lawrence SK, Weissman SM: Structure and nucleotide sequence of the heavy chain gene of HLA-DR.Proc Natl Acad Sci USA 1983;80:3543–3547.

31   Servenius B, Gustafsson K, Widmark E, Emmoth E, Andersson G, Larhammar D, Rask L, Peterson PA: Molecular map of the human HLA-SB(HLA-DP) region and sequence of an Sbalpha(DP-alpha) pseudogene. EMBO J 1984;3:3209–3214.

32   Van Endert PM, Lopez MT, Patel SD, Monaco JJ, McDevitt HO: Genomic polymorphism, recombination, and linkage disequilibrium in human major histocompatibility complex-encoded antigen-processing genes. Proc Natl Acad Sci USA 1992;89:11594–11597.

33   Baur MP, Danilovs JA: Reference tables for two and three locus haplotype frequencies of HLA-A, -B, -C, DR Bf and GLO; in Terasaki P (ed): Histocompatibility Testing. Los Angeles, UCLA Tissue Typing, 1980, pp 994–1210.

34   Hardy DA, Bell JI, Long EO, Lindsten T, McDevitt HO: Mapping of the class II region of the human major histocompatibility complex by pulsed-field gel electrophoresis. Nature 1986;323: 453–455.

35   Levine BB, Stember RH, Fontino M: Ragweed hayfever: Genetic control and linkage to HL-A haplotypes. Science 1972;178:1201–1203.

36   Marsh DG, Meyers DA, Bias WB: The epidemiology and genetics of atopic allergy. N Engl J Med 1981;305:1551–1559.

37   Blumenthal MN, Awdeh Z, Alper C, Yunis E: Ra5 immune responses, HLA antigens and complotypes (abstract). J Allergy Clin Immunol 1985;75:155.

38   Marsh DG, Blumenthal MN, Ishikawa T, Ruffilli A, Sparholt S, Freidhoff LR: HLA and specific immune responsiveness to allergens; in Tsuji K, Aizawa M, Sasazuki T (eds): HLA 1991. Proc Eleventh International Histocompatibility Workshop and Conference. Oxford, Oxford University Press, 1992, vol I, pp 765–767.

39   Marsh DG, Freidhoff LR, Ehrlich-Kautzky E, Bias WB, Roebber M: Immune responsiveness to *Ambrosia artemisiifolia* (short ragweed) pollen allergen *Amb a* VI (Ra6) is associated with HLA-DR5 in allergic humans. Immunogenetics 1987;26:230–236.

40   Young RP, Dekker JW, Wordsworth BP, Schou C, Pile KD, Matthiesen F, Rosenberg WMC, Bell JI, Hopkin JM, Cookson WOCM: HLA-DR and HLA-DP genotypes and immunoglobulin E responses to common major allergens. Clin Exp Allergy 1994;24:431–439.

41   Fischer GF, Pickl WF, Fae I, Ebner C, Ferreira F, Breiteneder H, Vikoukal E, Scheiner O, Kraft D: Association between IgE response against *Bet v* I, the major allergen of birch pollen, and HLA-DRB alleles. Hum Immunol 1992;33:259–265.

42   Sparholt SH, Georgsen J, Madsen HO, Svendsen UG, Schou C: Association between HLA-DRB3*0101 and immunoglobulin-E responsiveness to *Bet v* I. Hum Immunol 1994;39: 76–78.

43   Freidhoff LR, Ehrlich-Kautzky E, Meyers DA, Ansari AA, Bias WB, Marsh DG: Association of HLA-DR3 with human immune response to *Lol p* I and *Lol p* II allergens in allergic subjects. Tissue Antigens 1988;31:211–219.

44   Ansari AA, Freidhoff LR, Meyers DA, Bias WB, Marsh DG: Human immune responsiveness to *Lolium perenne* pollen allergen *Lol p* III (rye III) is associated with HLA-DR3 and DR5 [published erratum appears in Hum Immunol 1989;26:149]. Hum Immunol 1989;25:59–71.

45   Sriramarao P, Selvakumar B, Damodaran C, Rao BS, Prakash O, Rao PV: Immediate hypersensitivity to Parthenium hysterophorus. I. Association of HLA antigens and *Parthenium* rhinitis. Clin Exp Allergy 1990;20:555–560.

46   Cardaba B, Vilches C, Martin E, de Andres B, del Pozo V, Hernandez D, Gallardo S, Fernandez JC, Villalba M, Rodriguez R, Basomba A, Kreisler M, Palomino P, Lahoz C: DR7 and DQ2 are positively associated with immunoglobulin-E response to the main antigen of olive pollen (*Ole e* I) in allergic patients. Hum Immunol 1993;38:293–299.

47   Reid MJ, Nish WA, Whisman BA, Goetz DW, Hylander RD, Parker WA Jr, Freeman TM: HLA-DR4-associated nonresponsiveness to mountain cedar allergen. J Allergy Clin Immunol 1992;89:593–598.

48   Lympany P, Kemeny DM, Welsh KI, Lee TH: An HLA-associated nonresponsiveness to melittin: A component of bee venom. J Allergy Clin Immunol 1990;86:160–170.

49   Sasazuki T, Nishimura Y, Muto M, Ohta N: HLA-linked genes controlling immune response and disease susceptibility. Immunol Rev 1983;70:51–75.

50   Higgins JA, Thorpe CJ, Hayball JD, O'Hehir RE, Lamb J: Overlapping T-cell epitopes in the group I allergen of *Dermatophagoides* species restricted by HLA-DP and HLA-DR class II molecules. J Allergy Clin Immunol 1994;93:891–899.

51   Zeliszewski D, Gaudebout P, Golvano JJ, Dorval I, Prévost A, Borras-Cuesta F, Sterkers G: Molecular basis for degenerate T-cell recognition of one peptide in the context of several DR molecules. Hum Immunol 1994;41:28–33.

52   Brenner MB, Trowbridge IS, Strominger IL: Cross-linking of human T cell receptor proteins: Association between the T cell idiotype β subunit and the T3 glycoprotein heavy subunit. Cell 1985;40:183–190.

53   Hall C, Berkhout B, Alarcon B, Sancho J, Wileman T, Terhorst C: Requirements for cell surface expression of the human TCR/CD3 complex in non-T cells. Int Immunol 1991;3:359–368.

54   Carson GR, Kuestner RE, Ahmed A, Pettey CL, Concino MF: Six chains of the human T cell antigen receptor CD3 complex are necessary and sufficient for processing the receptor heterodimer to the cell surface. J Biol Chem 1991;266:7883–7887.

55   Change T-W, Testa D, Kung PC, Perry L, Dreskin HJ, Goldstein G: Cellular origin and interactions involved in the γ-interferon production induced by OKT3 monoclonal antibody. J Immunol 1982;128:585–589.

56   Wegener A-M, Letourneur F, Hoeveler A, Brocker T, Luton F, Malissen B: The T cell receptor/CD3 complex is composed of at least two autonomous transduction molecules. Cell 1992;68:83–95.

57   Rahemtulla A, Fung-Leung WP, Schilham MW, Kündig TM, Sambhara SR, Narendran A, Arabian A, Wakeham A, Paige CJ, Zinkernagel RM, Miller RG, Mak TW: Normal development and function of CD8 + cells but markedly decreased helper cell activity in mice lacking CD4. Nature 1991;353:180–184.

58   Fung-Leung W-P, Schilham MW, Rahemtulla A, Kündig TM, Vollenweider M, Potter J, van Ewijk W, Mak TW: CD8 is needed for development of cytotoxic T cells but not helper T cells. Cell 1991;65:443–449.

59   Littman DR: The structure of the CD4 and CD8 genes. Annu Rev Immunol 1987;5:561–584.

60   Gabert J, Langlet C, Zamoyska R, Parnes JR, Schmitt-Verhulst AM, Malissen B: Reconstitution of MHC class I specificity by transfer of the T cell receptor and Lyt-2 genes. Cell 1987;50:545–554.

61   Cobbold S, Qin S, Leong LYW, Martin G, Waldmann H: Reprogramming the immune system for peripheral tolerance with CD4 and CD8 monoclonal antibodies. Immunol Rev 1992;129:165–201.

62   Sleckman BP, Peterson A, Jones WK, Foran JA, Greenstein JL, Seed B, Burakoff SJ: Expression and function of CD4 in a murine T-cell hybridoma. Nature 1987;328:351–353.

63   Dembic Z, Haas W, Weiss S, McCubrey J, Kiefer H, von Boehmer H, Steinmetz M: Transfer of specificity by murine α and β T-cell receptor genes. Nature 1986;320:232–238.

64   Bank I, Chess L: Perturbation of the T4 molecule transmits a negative signal to T cells. J Exp Med 1985;162:1294–1303.

65   Wassmer P, Chan C, Lögdberg L, Shevach EM: Role of the L3T4 antigen in T cell activation. II. Inhibition of T cell activation by monoclonal anti-L3T4 antibodies in the absence of accessory cells. J Immunol 1985;135:2237–2242.

66   Owens T, de St Groth BF, Miller JFAP: Coaggregation of the T-cell receptor with CD4 and other T-cell surface molecules enhances T-cell activation. Proc Natl Acad Sci USA 1987;84:9209–9213.

67   Anderson P, Blue M-L, Morimoto C, Schlossman SF: Cross-linking of T3 (CD3) with T4 (CD4) enhances the proliferation of resting T lymphocytes. J Immunol 1987;139:678–682.

68   Rudd CE, Trevillyan JM, Dasgupta JD, Wong LL, Schlossman SF: The CD4 rectpor is complexed in detergent lysates to a protein-tyrosine kinase (pp58) from human T lymphocytes. Proc Natl Acad Sci USA 1988;85:5190–5194.

69   Veillette A, Bookman MA, Horak EM, Bolen JB: The CD4 and CD8 T cell surface antigens are associated with the internal membrane tyrosine-protein kinase p56[lck]. Cell 1988;55:301–308.

70   Saito H, Kranz DM, Takagaki Y, Hayday AC, Eisen HN, Tonegawa S: A third rearranged and expressed gene in a clone of cytotoxic T lymphocytes. Nature 1984;312:36–40.

71   Chien Y-H, Iwashima M, Kaplan KB, Elliott JF, Davis MM: A new T-cell receptor gene located within the alpha locus and expressed early in T-cell differentiation. Nature 1987;327:677–682.

72   Brenner MB, McLean J, Dialynas DP, Strominger JL, Smith JA, Owen FL, Seidman JG, Ip S, Rosen F, Krangel MS: Identification of a putative second T-cell receptor. Nature 1986;322:145–149.

73   Elliott JF, Rock EP, Patten PA, Davis MM, Chein Y-H: The adult T-cell receptor δ-chain is diverse and distinct from that of fetal thymocytes. Nature 1988;331:627–631.

74   Groh V, Porcelli S, Fabbi M, Lanier LL, Picker LJ, Anderson T, Warnke RA, Bhan AK, Strominger JL, Brenner MB: Human lymphocytes bearing T cell receptor γ/δ are phenotypically diverse and evenly distributed throughout the lymphoid system. J Exp Med 1989;169:1277–1294.

75   Inghirami G, Zhu BY, Chess L, Knowles DM: Flow cytometric and immunohistochemical characterization of the γ/δ lymphocyte population in normal human lymphoid tissue and peripheral blood. Am J Pathol 1990;136:357–367.

76   Janis EM, Kaufmann SHE, Schwartz RH, Pardoll DM: Activation of γδ T cells in the primary immune response to *Mycobacterium tuberculosis*. Science 1989;244:713–715.

77   O'Brien RL, Happ MP, Dallas A, Palmer E, Kubo R, Born WK: Stimulation of a major subset of lymphocytes expressing T cell receptor γδ by an antigen derived from *Mycobacterium tuberculosis*. Cell 1989;57:667–674.

78   Pfeffer K, Schoel B, Gulle H, Kaufmann SHE, Wagner H: Primary responses of human T cells to mycobacteria: A frequent set of γ/δ T cells are stimulated by protease resistant ligands. Eur J Immunol 1990;20:1175–1179.

79   Tsuji M, Mombaerts P, Lefrancois L, Nussenzweig RS, Zavala F, Tonegawa S: γδ T cells contribute to immunity against the liver stages of malaria in αβ T-cell deficient mice. Proc Natl Acad Sci USA 1994;91:345–349.

80   Janeway CA Jr, Jones B, Hayday AC: Specificity and function of cells bearing γδ T cell receptors. Immunol Today 1988;9:73–76.

81   Asarnow DM, Kuziel WA, Bonyhadi M, Tigelaar RE, Tucker PW, Allison JP: Limited diversity of γδ antigen receptor genes of Thy-1[+] dendritic epidermal cells. Cell 1988;55:837–847.

82   Allison JP, Havran WL: The immunobiology of T cells with invariant γδ antigen receptors. Annu Rev Immunol 1991;9:679–705.

83   Allison JP: Gamma delta T-cell development. Curr Open Immunol 1993;5:241–246.

84   Holt PG, McMenamin C: IgE and mucosal immunity: Studies on the role of intrapithelial Ia[+] dendritic cells and γ/δ T-lymphocytes in regulation of T-cell activation in the lung. Clin Exp Allergy 1991;21(suppl):148–152.

85   Janeway CA Jr: Molecular recognition: Frontiers of the immune system. Nature 1988;333:804–806.

86   McMenamin C, Pimm C, McKersey M, Holt PG: Regulation of IgE responses to inhaled antigen in mice by antigen-specific γδ T cells. Science 1994;265:1869–1871.

87 Caccia N, Bruns GAP, Kirsch IR, Hollis GF, Bertness V, Mak TW: T cell receptor α chain genes are located on chromosome 14 at 14q11-14q12 in humans. J Exp Med 1985;161:1255–1260.

88 Collins MKL, Goodfellow PN, Spurr NK, Solomon E, Tanigawa G, Tonegawa S, Owen MJ: The human T-cell receptor α-chain gene maps to chromosome 14. Nature 1985;314:273–274.

89 Loh EY, Lanier LL, Turck CW, Littman DR, Davis MM, Chien YH, Weiss A: Identification and sequence of a fourth human T cell antigen receptor chain. Nature 1987;330:569–372.

90 Caccia N, Kronenberg M, Saxe D, Haars R, Bruns GAP, Goverman J, Malissen M, Willard H, Yoshikai Y, Simon M, Hood L, Mak TW: The T cell receptor β chain genes are located on chromosome 6 in mice and chromosome 7 in humans. Cell 1984;37:1091–1099.

91 Murre C, Waldmann RA, Morton CC, Bongiovanni KF, Waldmann TA, Shows TB, Seidman JG: Human γ-chain genes are rearranged in leukaemic T cells and map to the short arm of chromosome 7. Nature 1986;316:549–552.

92 Davis MM, Bhorkman PJ: T-cell antigen receptor genes and T-cell recognition. Nature 1988;334: 395–402.

93 Clothia C, Boswell DR, Lesk AM: The outline structure of the T-cell αβ receptor. EMBO J 1988; 7:3745–3755.

94 Rosenberg WMC, Moss PAH, Bell JI: Variation in human T cell receptor V_β and J_β repertoire: Analysis using anchor polymerase chain reaction. Eur J Immunol 1992;22:541–549.

95 Moss PAH, Rosenberg WMC, Zintzaras E, Bell JI: Characterization of the human T cell receptor α-chain repertoire and demonstration of a genetic influence on V α usage. Eur J Immunol 1993; 23:1153–1159.

96 Gulwani-Akolkar B, Posnett DN, Janson CH, Grunewald J, Wigzell H, Akolkar P, Gregersen PK, Silver J: T cell receptor V-segment frequencies in peripheral blood T cells correlate with human leukocyte antigen type. J Exp Med 1991;174:1139–1146.

97 Akolkar PN, Gulwani-Akolkar B, Pergolizzi R, Bigler RD, Silver J: Influence of HLA genes on T cell receptor V segment frequencies and expression levels in peripheral blood lymphocytes. J Immunol 1993;150:2761–2773.

98 Gulwani-Akolkar B, Shi B, Akolkar PN, Ito K, Bias WB, Silver J: Do HLA genes play a prominent role in determining T cell receptor Vα segment usage in humans? J Immunol 1995;154: 3843–3851.

99 Vissinga CS, Charmley P, Concannon P: Influence of coding region polymorphism on the peripheral expression of a human TCR Vβ gene. J Immunol 1994;152:1222–1227.

100 Donahue JP, Ricalton NS, Behrendt CE, Rittershaus C, Calaman S, Marrack P, Kappler JW, Kotzin BL: Genetic analysis of low Vβ3 expression in humans. J Exp Med 1994;179:1701–1706.

101 Maksymowych WP, Gabriel CA, Luyrink L, Melin-Aldana H, Elma M, Giannini EH, Lovell DJ, Van Kerckhove C, Leiden J, Choi E, Glass DN: Polymorphism in a T-cell receptor variable gene is associated with susceptibility to a juvenile rheumatoid arthritis subset. Immunogenetics 1992;35: 257–262.

102 Tebib JG: Alcocer-Varela J, Alarcon-Segovia D, Schur PH: Association between a T-cell receptor restriction fragment length polymorphism and systemic lupus erythematosus. J Clin Invest 1990; 86:1961–1967.

103 Kumar V, Kono DH, Urban JL, Hood L: The T cell receptor repertoire and autoimmune diseases. Annu Rev Immunol 1989;7:657–682.

104 Concannon P, Wright JA, Wright LG, Sylvester DR, Spielman RS: T-cell receptor genes and insulin-dependent diabetes mellitus: No evidence for linkage from affected sib pairs. Am J Hum Genet 1990;47:45–52.

105 Posnett DN: Allelic variations of human TCR V gene products. Immunol Today 1990;11:368–373.

106 Concannon P, Gatti RA, Hood LE: Human T cell receptor Vβ gene polymorphism. J Exp Med 1987;165:1130–1140.

107 Li Y, Szabo P, Robinson MA, Dong B, Posnett DN: Allelic variations in the human T cell receptor Vβ6.7 gene products. J Exp Med 1990;171:221–230.

108 Robinson MA: Allelic sequence variations in the hypervariable region of a T-cell receptor β chain: Correlation with restriction fragment length polymorphism in human families and populations. Proc Natl Acad Sci USA 1989;86:9422–9426.

109 Plaza A, Kono DH, Theofilopoulos AN: New human Vβ genes and polymorphic variants. J Immunol 1991;147:4360–4365.

110 Cornélis F, Pile K, Loveridge J, Moss P, Harding R, Julier C, Bell J: Systematic study of human αβ T cell receptor V segments shows allelic variations resulting in a large number of distinct T cell receptor haplotypes. Eur J Immunol 1993;23:1277–1283.

111 Hansen T, Ronningen KS, Ploski R, Kimura A, Thorsby E: Coding region polymorphisms of human T-cell receptor V beta 6.9 and V beta 21.4. Scand J Immunol 1992;36:285–290.

112 Gomolka M, Epplen C, Buitkamp J, Epplen JT: Novel members and germline polymorphisms in the human T-cell receptor Vb6 family. Immunogenetics 1993;37:257–265.

113 So A, John S, Bailey C, Owen MJ: A new polymorphic marker of the T-cell antigen receptor α chain genes in man. Immunogenetics 1987;25:141–144.

114 Wright JA, Hood L, Cancannon P: Human T-cell receptor Vα gene polymorphism. Hum Immunol 1991;32:277–283.

115 Reyburn H, Cornélis F, Russell V, Harding R, Moss P, Bell J: Allelic polymorphism of human T cell receptor V alpha gene segments. Immunogenetics 1993;38:287–291.

116 Charmley P, Nickerson D, Hood L: Polymorphism detection and sequence analysis of human T-cell receptor Vα-chain-encoding gene segments. Immunogenetics 1994;39:138–145.

117 Ibberson MR, Copier JP, So AK: Genomic organization of the human t-cell receptor variable α gene cluster. Genomics 1995;28:131–139.

118 Seboun E, Robinson MA, Kindt TJ, Hauser SL: Insertion/deletion-related polymorphisms in the human T cell receptor β gene complex. J Exp Med 1989;170:1263–1270.

119 Foote J, Winter G: Antibody framework residues affecting the conformation of the hypervariable loops. J Mol Biol 1992;224:487–499.

120 Saito T, Germain RN: Marked differences in the efficiency of expression of distinct αβ T cell heterodimers. J Immunol 1989;143:3379–3384.

121 Fink PJ, Matis LA, McElligott DL, Bookman M, Hedrick SM: Correlations between T-cell specificity and the structure of the antigen receptor. Nature 1986;321:219–226.

122 Winoto A, Urban JL, Lan NC, Goverman J, Hood L, Hansberg D: Predominant use of a Vα gene segment in mouse T-cell receptors for cytochrome c. Nature 1986;324:679–682.

123 Loveridge JA, Rosenberg WMC, Kirkwood TBL, Bell JI: The genetic contribution to human T-cell receptor repertoire. Immunology 1991;74:246–520.

124 Malhotra U, Spielman R, Concannon P: Variability in T cell receptor V β gene usage in human peripheral blood lymphocytes: Studies of identical twins, siblings,and insulin-dependent diabetes mellitus patients. J Immunol 1992;149:1802–1808.

125 Achta-Orbea H, Mitchell DJ, Timmermann L, Wraith DC, Tausch GS, Waldor MK, Zamvil SS, McDevitt HO, Steinman L: Limited heterogeneity of T cell receptors from lymphocytes mediating autoimmune encephalomyelitis allows specific immune intervention. Cell 1988;54:263–273.

126 Vandenbark AA, Hashim G, Offner H: Immunization with a synthetic T-cell receptor V-region peptide protects against experimental autoimmune encephalomyelitis. Nature 1989;341:541–544.

127 Giegerich G, Pette M, Meinl E, Epplen JT, Wekerle H, Hinkkanen A: Diversity of T cell receptor α and β chain genes expressed by human T cells specific for similar myelin basic protein peptide/ major histocompatibility complexes. Eur J Immunol 1992;22:753–758.

128 Utz U, Biddison WE, McFarland HF, McFarlin DE, Flerlage M, Martin R: Skewed T-cell receptor repertoire in genetically identical twins correlates with multiple sclerosis. Nature 1993;364:243–247.

129 Hedrick SM, Engel I, McElligott DL, Fink PJ, Hsu M-L, Hansburg D, Matis LA: Selection of amino acid sequences in the beta chain of the T cell antigen receptor. Science 1988;239:1541–1544.

130 Sorger SB, Hedrick SM: Highly conserved T-cell receptor junctional regions. Evidence for selection at the protein and the DNA level. Immunogenetics 1990;31:118–122.

131 Danska JS, Livingstone AM, Paragas V, Ishihara T, Fathman CG: The presumptive CDR3 regions of both T cell receptor α and β chains determine T cell specificity for myoglobin peptides. J Exp Med 1990;172:27–33.

132 Moss PAH, Moots RJ, Rosenberg WMC, Rowland-Jones SJ, Bodmer HC, McMichael AJ, Bell JI: Extensive conservation of α and β chains of the human T-cell antigen receptor recognizing HLA-A2 and influenza A matrix peptide. Proc Natl Acad Sci USA 1991;88:8987–8990.

133   Boitel B, Ermonval M, Panina-Bordignon P, Mariuzza RA, Lanzavecchia A, Acuto O: Preferential Vβ gene usage and lack of junctional sequence conservation among human T cell receptors specific for a tetanus toxin-derived peptide: Evidence for a dominant role of a germline-encoded V region in antigen/major histocompatibility complex recognition. J Exp Med 1992;175:765–777.

134   Henwood J, Loveridge J, Bell JI, Hill Gaston JS: Restricted T cell receptor expression by human T cell clones specific for mycobacterial 65-kDa heat-shock protein: Selective in vivo expansion of T cells bearing defined receptors. Eur J Immunol 1993;23:1256–1265.

135   Sottini A, Imberti L, Gorla R, Cattaneo R, Primi D: Restricted expression of T cell receptor Vβ but not Vα genes in rheumatoid arthritis. Eur J Immunol 1991;21:461–466.

136   Davies TF, Martin A, Concepcion ES, Graves P, Cohen L, Ben-Nun A: Evidence of limited variability of antigen receptors on intrathyroidal T cells in autoimmune thyroid disease. N Engl J Med 1991;325:238–244.

137   Grom AA, Thompson SD, Luyrink L, Passo M, Choi E, Glass DN: Dominant T-cell-receptor β chain variable region V β 14⁺ clones in juvenile rheumatoid arthritis. Proc Natl Acad Sci USA 1993;90:11104–11108.

138   Pluschke G, Ricken G, Taube H, Kroninger S, Melchers I, Peter HH, Eichmann K, Krawinkel U: Biased T cell receptor Vα region repertoire in the synovial fluid of rheumatoid arthritis patients. Eur J Immunol 1991;21:2749–2754.

139   Uematsu Y, Wege H, Straus A, Ott M, Bannwarth W, Lanchbury J, Panayi G, Steinmetz M: The T-cell receptor repertoire in the synovial fluid of a patient with rheumatoid arthritis is polyclonal. Proc Natl Acad Sci USA 1991;88:8534–8538.

140   McIntosh RS, Tandon N, Pickerill AP, Davies R, Barnett D, Weetman AP: The γδ T cell repertoire in Graves' disease and multinodular goitre. Clin Exp Immunol 1993;94:473–477.

141   McIntosh RS, Watson PF, Pickerill AP, Davies R, Weetman AP: No restriction of intrathyroidal T cell receptor V α families in the thyroid of Graves' disease. Clin Exp Immunol 1993;91:147–152.

142   Wedderburn LR, O'Hehir RE, Hewitt CRA, Lamb JR, Owen MJ: In vivo clonal dominance and limited T-cell receptor usage in human CD4+ T-cell recognition of house dust mite allergens. Proc Natl Acad Sci USA 1993;90:8214–8218.

143   Mohapatra SS, Mohapatra S, Yang M, Ansari AA, Parronchi P, Maggi E, Romagnani S: Molecular basis of cross-reactivity among allergen-specific human T cells: T-cell receptor Vα gene usage and epitope structure. Immunology 1994;81:15–20.

144   Breiteneder H, Scheiner O, Hajek R, Hulla W, Hüttinger R, Fischer G, Kraft D, Ebner C: Diversity of TCRAV and TCRBV seqences used by human T-cell clones specific for a minimal epitope of Bet v I, the major birch pollen allergen. Immunogenetics 1995;42:53–58.

145   Huang SK, Zwollo P, Marsh PG: Class II major histocompatibility complex restriction of human T cell responses to short ragweed allergen, Amb a V. Eur J Immunol 1991;21:1469–1473.

146   Huang SK, Yi M, Palmer E, Marsh DG: A dominant T cell receptor beta-chain in response to a short ragweed allergen, Amb a V. J Immunol 1995;154:6157–6162.

147   Renz H, Saloga J, Bradley KL, Loader JE, Greenstein JL, Larsen G, Gelfand EW: Specific Vβ T cell subsets mediate the immediate hypersensitivity response to ragweed allergen. J Immunol 1993;151:1907–1917.

148   Schiffer M, Kabat EA, Wu TT: Subgroups of TCR α chains and correlation with T-cell function. Immunogenetics 1992;35:224-234.

149   Cihak J, Hoffmann-Fezer G, Ziegler-Heibrock HWL, Stein H, Kaspers B, Chen CH, Cooper MD, Lösch U: T cells expressing the Vβ1 T-cell receptor are required for IgA production in the chicken. Proc Natl Acad Sci USA 1991;88:10951–10955.

150   Renz H, Bradley KL, Marrack P, Gelfand EW: T cells expressing variable elements of T-cell receptor β8 and β2 chain regulate murine IgE production. Proc Natl Acad Sci USA 1992;89:6438–6442.

151   Renz H, Bradley K, Saloga J, Loader J, Larsen GL, Gelfand EW: T cells expressing specific Vβ elements regulate immunoglobulin E production and airways responsiveness in vivo. J Exp Med 1993;177:1175–1180.

152   Loh EY, Elliott JF, Cwirla S, Lanier LL, Davis MM: Polymerase chain reaction with single-sided specificity: Analysis of T cell receptor δ chain. Science 1989;243:217–220.

153 Uematsu Y: A novel and rapid cloning method for the T cell receptor variable region sequences. Immunogenetics 1991;34:174–178.

154 Choi Y, Kotzin B, Herron L, Callahan J, Marrack P, Kappler J: Interaction of *Staphylococcus aureus* toxin 'superantigen' with human T cells. Proc Natl Acad Sci USA 1989;86:8941–8945.

155 Seguardo OG, Schendel DJ: Identification of predominant T-cell receptor rearrangements by temperature-gradient gel electrophoresis and automated DNA sequencing. Electrophoresis 1993;14:747–752.

156 Cornélis F, Hashimoto L, Loveridge J, MacCarthy A, Buckle V, Bell J: Identification of a CA repeat at the TCRA locus using yeast artificial chromosomes: A general method for generating highly polymorphic markers at chosen loci. Genomics 1992;13:820–825.

157 Moffatt MF, Hill MR, Cornélis F, Schou C, Faux JA, Young RP, James AL, Ryan G, le Souef P, Musk AW, Hopkin JM, Cookson WOCM: Genetic linkage of T-cell receptor α/δ complex to specific IgE responses. Lancet 1994;343:1597–1600.

158 Moffatt MF: Genetic Studies of Atopy; DPhil thesis, Bodleian Library, Oxford 1993.

159 Deichmann KA, Hildebrandt F, Kuehr J, Forster J: Genetic linkage analysis of predicted asthma genes and atopy. Allergy 1995;50(suppl):164.

160 Moffatt MF, Young A, Schou C, Faux JA, Musk AW, Cookson WOCM: Involvement of TCR-α/δ and HLA-DR in specific allergy. Allergy 1995;50(suppl):164.

161 Charpin D, Birnbaum J, Haddi E, Genard G, Lanteaume A, Toumi M, Faraj F, Van der Brempt X, Vervolet D: Altitude and allergy to house dust mites. A paradigm of the influence of environmental exposure on allergic sensitisation. Am Rev Respir Dis 1991;143:983–986.

162 Holt PG, McMenamin C, Nelson D: Primary sensitisation to inhalant allergens during infancy. Pediatr Allergy Immunol 1990;1:3–15.

163 Sporik R, Holgate S, Platts-Mills TAE, Cogswells JJ: Exposure to house dust mite allergen der P1 and the development of asthma in children. N Engl J Med 1990;323:502–507.

164 Von Mutius E, Fritzsch C, Weiland SK, Roell G, Magnussen H: Prevalence of asthma and allergic disorders among children in united Germany: A descriptive comparison. Br Med J 1992;305:1395–1399.

165 Von Mutius E, Martinez FD, Fritzsch C, Nicolai T, Roell G, Thiemann HH: Prevalence of asthma and atopy in two areas of West and East Germany. Am J Respir Crit Care Med 1994;149:358–364.

166 Bråbäck L, Breborowicz A, Dreborg S, Knutsson A, Pieklik H, Björkstén B: Atopic sensitization and respiratory symptoms among Polish and Swedish school children. Clin Exp Allergy 1994;24:826–835.

167 Von Mutius E, Martinez FD, Fritzsch C, Nicolai T, Reitmer P, Thiemann HH: Skin test reactivity and number of siblings. Br Med J 1994;308:692–695.

168 Lynch NR, Hagel I, Perez M, Di Prisco MC, Lopez R, Alvarez N: Effect of antihelminthic treatment on the allergic reactivity of children in a tropical slum. J Allergy Clin Immunol 1993;92:404–411.

169 Paré PD, Faux JA, Hill MR, Kan R, Bremmer P, Musk M, Murray C, le Souef PN, Musk AW, Cookson WOCM: Skin test responses and specific IgE levels to common aeroallergens in Australian aborigines. Am J Resp Crit Care Med 1995;151:A214.

Miriam F. Moffatt, Asthma Genetics Group, Nuffield Department of Clinical Medicine, John Radcliffe Hospital, Oxford OX3 9DU (UK)

Hall, IP (ed): Genetics of Asthma and Atopy.
Monogr Allergy. Basel, Karger, 1996, vol 33, pp 97–108

....

Molecular Genetics of the High-Affinity IgE Receptor

J. M. Hopkin

Osler Chest Unit, Churchill Hospital, Oxford, UK

Atopy is a heterogeneous disorder in which genetic and environmental factors interact to cause enhanced IgE responses and the clinical allergy that is typical of asthma, hay fever and childhood eczema.

The high-affinity IgE receptor, bound to maternal mast cells and other cells, plays a central role in IgE-mediated allergic disorders. This paper presents evidence that the β-subunit of *FcεRI*, encoded on chromosome 11, is one important source of genetic variants that underlie the atopy syndrome.

Atopy and Its Origins

Atopy is a state of allergic response to common, largely innocuous environmental antigens, such as those derived from house dust mites and pollen grains. The allergic reaction involves many cells of the immune system and their diverse molecular products; characteristic features of the response are the exuberant production of immunoglobulin E [1], and the binding of this to mucosal mast cells via the high-affinity IgE receptor [1, 2]. This immunological disorder causes intense mucosal inflammation that affects the bronchus, nose and skin in variable combination, which results in the clinical disorders of asthma, rhinitis (hay fever or pollinosis) and dermatitis (eczema) [3].

Atopy is multifactorial in origin and results from the interaction of a variety of genetic and environmental factors. Most atopics are allergic to more than one common environmental antigen and the causes of atopy probably act at a variety of levels: (a) generalized hyper-IgE responsiveness; (b) IgE response to specific allergens or epitopes; (c) end-organ expression. It is likely that the predominant influences at each level differ.

There are clear indicators of strong *environmental factors*. A priori, there would be no atopy in an environment with minimal allergen quantities. Studies on populations of similar ethnicity but varying culture/lifestyle show great disparities in the prevalence of atopy [4]. Westernization per se is associated with an increased risk of atopic disorder of some 2- to 3-fold [5, 7]. Migrant studies support the cultural/lifestyle effect since immigrants rapidly acquire the risk of atopy of their new home, e.g. Tokelauans in New Zealand [8] and Asians and West Indians in Britain [9]. Within single communities, increasing prevalence and severity of atopic clinical disorders have been noted in the past 20–30 years [10]. Critical analysis of these trends suggest that increased awareness and changes in diagnostic habits do not account for the changes. These changes in prevalence of atopy across cultural divides and over short spans of time, indicate that environmental changes (associated with Westernization) have a profound impact on the risk of disease. These urgently require discovery.

The plainest evidence of significant *genetic effects* is derived from twin studies. All of those point to an important genetic effect [11, 13] but a recent American study, on monzygotic and dizygotic twins reared together and apart, provides the most informative and reliable data [14]. Whether reared together or apart, monozygous twins showed significantly greater correlation coefficients for total serum IgE levels; these indicated that some 50% of the variation in total IgE levels can be attributed to genetic factors. Also, the monozygous twins showed 71% concordance for IgE response to *any* one or more common allergens; the discordance of 29% is a reminder of significant environmental interaction. Therefore, twin studies indicate that there are important genetic influences on the phenotype of generalized IgE responsiveness. But the genetic effects on allergen-specific sensitization and clinical disorder seem significantly weaker, e.g. concordance for sensitization to grass pollen (*Phleum pretense*) was approximately 15% in dizygotic twins and less than 40% for monozygotic twins [14]. Many studies on the familial aggregation of atopy, but which offer less rigorous genetic interpretation, support these conclusions.

These various studies show that atopic disorder is due to the interaction of genetic and environmental factors but do not indicate how many factors there are and how interactive these may be. The outbred nature of man and the range of his culture and environment, suggest that heterogeneity is also likely; therefore the factors underlying the phenotype in different individuals will differ significantly. There must also be cases where a significant predisposing genetic variable is present but insufficient environmental interaction to cause the phenotype (variable genetic penetrance), and others where substantial environmental factors may cause the atopy phenotype even when genetic factors present are not important (phenocopies).

Approaches to Genetic Analysis

The multifactorial and heterogeneous origin of atopy and variable pene-trance and phenocopies creating diagnostic errors, greatly complicates the approach to investigating its genetics. But, the power of molecular genetic methods, allied to large-scale family resources, does offer the potential to identify an unknown number of genetic loci where variants can significantly influence the risk of disorder [15, 16]. Their identification might lead to the development of predictive genetic assays and of pharmacological intervention targeted on the 'variant' molecular behaviour. Indeed the use of particular targeted pharmacological treatments might depend on an individual's inherited molecular pathology.

Molecular Biology and Candidate Genes

The advent of the interweaving techniques and concepts of modern mo-lecular and cell biology has allowed the recognition of a range of 'pro-allergic/inflammatory' protein products relevant to the atopic disease process; their roles in various cells including mast cells, eosinophils and lymphocyte subtypes [2, 17, 18] are being clarified and for many, their genes have been cloned and sequenced.

Each – for example the interleukins IL-4 and IL-13 or their respective receptors which promote switching from IgM to IgE in the B lymphocyte [19] or FcεRI (or its subunits) – might offer a 'candidate gene', for atopy in which it is suspected that there might be variants/mutants within the gene or its controlling elements, which cause quantitative or qualitative changes in its protein and thereby risk of atopy.

When the genetic sequence is known, variants can now be readily sought by molecular analysis; then their importance can be tested either by functional assays (tortuous and difficult) or by epidemiological surveys of association between the variants and disease in the general population (more efficient). In either event, the list of potential 'candidate' genes is very long and therefore promoting the cause of one in comparison with another, as a case for investiga-tion, must be viewed as 'best guessing'.

Positional Cloning

The 'reverse genetics' or 'positional cloning' approach evidently offers a more unbiased approach, whose functioning has been greatly enhanced by the recent identification of very large numbers of polymorphic, highly informative, markers spaced at appropriately short intervals along the human chromosomes [16, 20, 23]. The exclusive use of variable dinucleotide repeats, examined by automated DNA amplification and electrophoresis, has led to successful delineation of the major genetic effects underlying type I diabetes [15]. The approach is also perfectly valid for atopy; it is based on the simple principle that 'atopy loci', will be identified by coinheritance of disease with defined genetic markers, because of their chromosomal proximity (genetic linkage). Establishing a chromosomal location is followed by more defined genetic linkage mapping, using a larger number of polymorphic genetic markers within the defined chromosomal region, before testing structural genes at the site, previously known or unknown, as candidates [24]. Definition of variants, within the coding sequence or controlling elements of a structural gene, and the demonstration of association with the disease in a population survey is the immediate goal [25]; it is the basis on which functional studies are then conducted [26, 27]. However, heterogeneity, variable penetrance and phenocopy effects all hamper fine linkage mapping in atopy; amalgamation of information from chromosomal linkages with the wealth of molecular biological now available on atopy is a counterbalancing advantage.

The complexity of atopy (a common heterogeneous disorder, with undefined modes of inheritance, and variable penetrance) makes admixture effects highly likely in large rambling families and makes standard genetic linkage analysis, by lod scores, very difficult and potentially misleading [28]. The method of affected sibling-pairs [29, 30] offers substantial advantages under these circumstances because mode(s) of inheritance does not need to be assumed, and because it minimizes admixture and variable penetrance effects. Analysis for genetic linkage can be reduced to a simple test of whether atopy affected siblings share parental alleles at particular loci beyond the expected or random proportions of 50% from any single parent for a dominant genetic effect, or 25% from both parents for a recessive effect.

Linkage of Atopy to Chromosome 11

Using quantitative assays for IgE response to allergens, we observed genetic linkage between generalized atopic IgE responses and chromosome 11q. Originally found in a small set of extended families and using lod score statistics, we extended our observations to a data set which included over 300 affected sibling-pairs [31–34]. The observation of linkage is robust to phenotype classification [35]. The data suggested that some 60% of families, ascertained through a young symptomatic atopic proband from Southern England, were linked to chromosome 11q [35].

Remarkably, the sharing of alleles from chromosome 11 by atopic sibling-pairs was exclusively from maternal chromosomes [33]. This observation accords with data from large epidemiological studies suggesting a maternal transmission of atopy [36–38]. It is consistent either with a maternal effect on fetal or neonatal immune development or with paternal genomic imprinting.

This latter genetic mechanism is being increasingly recognized and in it there is parent specific inactivation of certain genes over one generation.

Early attempts at independent replication of linkage to chromosome 11q produced variable results. Genetic heterogeneity and methodological factors, in particular the numbers of families and individuals tested, probably account for the discrepancies. Four studies report negative linkage [39–42]. Two studies contained insufficient information to exclude linkage of atopy to the marker *D11S97* on chromosome 11 [39, 40]. Inspection of the raw data from a third study [41] of three extended pedigrees shows a maximum lod score of 1.7 at 0 recombination in one family; the other two families show paternal inheritance and nonlinkage of atopy. A further study, of mixed extended and nuclear families, tested linkage with the locus *Int2* which is telomeric to *D11S97*, although atopy had previously been reported as 10% centromeric to the marker; the lod score was –2 at 10% recombination [42]. None of these studies took account of the maternal linkage to chromosome 11. Data from Japan, using lod scores [43], and the Netherlands, using affected sib-pair methods [44], now confirm linkage in families with marked symptomatic atopy.

FcεRI-β

In linkage mapping of atopy on chromosome 11q we defined a confidence interval for the localization of the atopy locus around 2 homologous genes, *CD20* and the *β-subunit of FcεRI* [45]. *CD20* is a proliferation and differentiation factor in B-lymphocyte lineage [46] whose function is not known to be related to atopic IgE responses. We found that *CD20 MspI* restriction alleles are not associated with atopy in children from unrelated nuclear families (odds ratio for alleles A and B = 0.95, 95% CI 0.56–1.60).

But *FcεRI-β* seemed a far more interesting prospect for the atopy locus on chromosome 11q. *FcεRI* is comprised of three subunits α, β and $γ_2$ [47, 48]; in man α and γ are encoded on chromosome 1 and the β subunit on chromosome 11. *FcεRI* is expressed on mast cells, basophils, monocytes and Langerhans' cells [49, 50]. The receptor plays a central role in the mediation of IgE-dependent allergic inflammation [51, 52] but also in IgE metabolism and mast cell and B-lymphocyte differentiation and growth [53, 54]. Stimulation of *FcεRI* causes release from the mast cell of cytokines, including IL-4, which are implicated in the up-regulation of mast cell and helper T-cell subtype 2 (TH2) development and of IgE production by β lymphocytes [55–58]. Lung mast cells that express cell contact signals including CD40 ligand may, in the presence of IL-4, regulate local B-lymphocyte IgE production independently of T lymphocytes [59]. Variants of *FcεRI-β* might promote the atopic state either by enhanced release of pro-inflammatory mediators by mast cells (to cause more

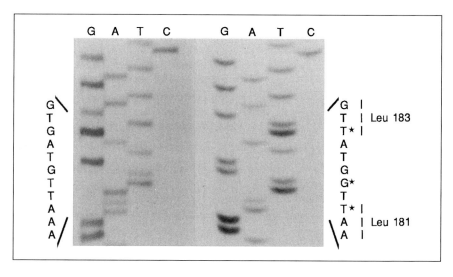

Fig. 1. DNA sequence at the junction of the 5th intron and 6th exon of *FcεRI-β* from a nonatopic and atopic – the latter showing three changes predicting the substitution of leucines at amino acid positions 181 and 183.

symptomatic disease) or by enhanced mast cell expression of IL-4 and CD40 ligand (to cause more local B-lymphocyte IgE production).

We therefore sequenced genomic DNA from each of the 7 exons and splice sites of *FcεRI-β* in 6 atopic and 6 nonatopic individuals. One atopic individual was found to have a chromosome with 3 nucleotide substitutions in the 6th exon (fig. 1). These substitutions predicted the substitution of leucine for isoleucine (position 181) and leucine for valine (position 183) within the 4th transmembrane domain of Fcε-β [60].

The prevalence of leucine residues at 181 and 183 of *FcεRI-β* and their relationship to atopy were defined using allele-specific DNA amplification (ARMS) [61] (fig. 2). Two subject groups were studied: a random sample of patients (unselected for atopy) having full blood counts performed and 60 nuclear families ascertained through a young atopic asthmatic proband (age 5–25 years) with at least one other sibling and both parents (regardless of atopy status) available for study.

In the random patient sample, Leu 181 was found in 25 of 163 individuals (15%) of whom one was homozygous; none showed a Leu 183 substitution. Associations were found between the presence of Leu 181 and high total serum IgE (odds ratio (OR) 3.07 (95% confidence interval 1.25–7.55, Fisher's statistic (FS) = 5.96, p = 0.01)) and positive IgE tests to grass pollen antigen (OR 2.61

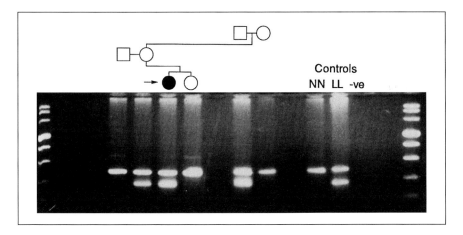

Fig. 2. Testing for the Leu 181 mutation (the lower band shown) by allele-specific DNA amplification (ARMS). In this family an atopic girl (solid symbol) inherits Leu 181 from a maternal grandfather and a nonatopic mother.

(95% CI 1.07–6.4), FS 4.48, p = 0.03) but not to house dust mite antigen (OR 1.44, 95% CI 0.6–3.5). Thirteen (56%) of the Leu 181 positive subjects were designated atopic (12 by positive RAST tests) and showed a mean total serum IgE of 300 kU/l; total serum IgE varies with age, race and other variables but the upper limit of normal, by association with allergen sensitization and allergic symptoms, is estimated to be about 100 kU/l in nonsmoking adults in Western populations.

The results from the 60 nuclear families were as follows: Ten (17%) of the families were found to have the Leu 181 variant segregating; this was confirmed by DNA sequencing. In each family the Leu 181 was maternally inherited (FS = 22.2, p < 0.0001). Amongst the children Leu 181 showed a strong association with atopy (all 12 children with Leu 181 were atopic, 10 of 12 Leu 181-negative children were not nonatopic, FS = 18.4, p < 0.0001). Atopy was observed in a child without Leu 181 in families 2 and 10 and in each instance the father also had atopy without Leu 181. Eight of the 10 Leu 181 heterozygous mothers were themselves atopic. DNA was available from both maternal grandparents in two families; Leu 181 was of grandmaternal origin where the Leu 181 mother was atopic and of grandpaternal origin where the Leu 181 mother was nonatopic. Inheritance of Leu 181 from a mother is highly predictive of atopy in these ten families, all 13 such individuals were atopic.

The phenotype in these family subjects was of marked atopy. Only 2 of 14 atopic children showed elevation of total IgE without allergen-specific responses and many of the probands had hay fever and eczema in addition to asthma.

The strong association between maternally inherited Leu 181 and atopy in a set of unrelated families indicate that variants of *FcεRI-β* may be one cause of atopic IgE responsiveness. This possibility is consistent with the known biological functions of the high-affinity IgE receptor [2].

Our data cannot determine whether Leu 181, within the 4th TM of *FcεRI-β*, alters receptor function or whether it is in linkage disequilibrium with undiscovered functional variants of *FcεRI-β* or its controlling elements. In the atopic subject originally identified as Leu 181 and Leu 183, no other mutation was detected in full coding and splice site sequences of *FcεRI-β*. Alpha helical TMs play an important role in the function *FcεRI* [62] and other similar receptors [63] in which nonionic interactions between nonpolar amino acids regulate the relationship of the helices and influence signal transduction [64]. Mutagenicity studies on the *FcεRI* subunits show that amino acid substitution in TMs can cause significant changes in the receptor's expression and function [65]. Single amino acid changes within TMs of other seven-helix bundle receptors have major functional effects; these include 10 to 20-fold changes in ligand binding in the 5-hydroxytryptamine receptor [66], and causation of familial male precocious puberty in the luteinizing hormone receptor [67]. The exchanges of aliphatic amino acids (Ise-Val-Leu) within a TM of *FcεRI-β* parallel species-specific variants of the brain cholecystokinin-B/gastrin receptor which result in 20-fold altered affinity for benzodiazepine-based antagonists [68]. It may be significant that substitution of leucines at positions 181 and 183 in human *FcεRI-β* generates the same sequence documented in rodents [69, 70].

Conclusion

The data suggest that variants of *FcεRI-β* on chromosome 11 are one significant genetic cause of atopy. But atopy is clearly heterogeneous. We have tested whether there might be, analogously to thalassaemia, genetic linkage between atopy and polymorphisms closely related to the α and γ subunits of *FcεRI* on chromosome 1 also. We have found none. We have however observed genetic linkage between IgE response to allergen and a microsatellite polymorphism closely related to the α/δ complex of the T-cell receptor on chromosome 14 [71]. Marsh et al. [72] have reported linkage between total serum IgE levels and chromosome 5 in Amish families. Of note, this site on chromosome

5 includes a series of genes for cytokines, including IL-4, the prime promoter of IgE class-switching. Comprehensive genome searches for linkages to atopy are currently underway and will clarify the extent of genetic heterogeneity and the relative importance of the various loci in different populations.

References

1   Ishizaka K: Maechanisms of reaginic hypersensitivity. Clin Allergy 1971;1:9–24.
2   Metzger H: The high-affinity receptor for IgE on mast cells. Clin Exp Allergy 1991;21:269–279.
3   Stenius B, Wide L, Seymour WM, Holford-Strevens V, Pepys J: Clinical significance of specific IgE to common allergens. I. Relationship of specific IgE against *Dermatophagoides* spp. and grass pollen to skin and nasal tests and history. Clin Allergy 1971;1:37–55.
4   Leung R, HoP: Asthma, allergy, and atopy in three south-east Asian populations. Thorax 1994; 49:1205–1210.
5   Peat JK, Haby M, Spijker J, Berry G, Woolcock AJ: Prevalence of asthma in adults in Busselton, Western Australia. BMJ 1992;305:1326–1329.
6   Zhong NS, Chen RC, O'Yang M: Bronchial hyperresponsiveness in young students of southern China: Relation to respiratory symptoms, diagnosed asthma, and risk factors. Thorax 1991;45: 860–865.
7   Leung R, Tseng RyM: Allergic diseases in Hong Kong schoolchildren. Hong Kong Pract 1993;15: 2409–2420.
8   Waite DA, Eyles EF, Tonkin SL, O'Donnell TV: Asthma prevalence in Tokelauan children in two environments. Clin Allergy 1980;10:71–75.
9   Morrison-Smith J, Cooper S: Asthma and atopic disease in immigrants from Asia and the West Indies. Postgrad Med J 1981;57:774–776.
10   Bruce IN, Harland RW, McBride NA, MacMahon J: Trends in the prevalence of asthma and dyspnoea in first year university students, 1972–89. QJ Med 1993;86:425–430.
11   Lubs ME: Allergy in 7,000 twin pairs. Acta Allergol 1974;26:249–285.
12   König P, Godfrey S: Exercise-induced bronchial liability in monozygotic (identical) and dizygotic (nonidentical) twins. J Allergy Clin Immunol 1974;54:280–287.
13   Hopp RJ, Bewtra AK, Watt GD, Nair NM, Townley RG: Genetic analysis of allergic disease in twins. J Clin Immunol 1984;73:265–270.
14   Hanson B, McGue M, Roitman-Johnson B, Segal NL, Bouchard TJ Jr, Blumenthal MN: Atopic disease and immunoglobulin E in twins reared apart and together. Am J Hum Genet 1991;48:873–879.
15   Davies JL, Kawaguchi Y, Bennett ST, Copeman JB, Cordell HJ, Pritchard Le, Reed PW, Reed PW, Gough SC, Jenkins SC, Palmer SM: A genome-wide search for human type I diabetes susceptibility genes. Nature 1994;371:130–136.
16   Little P: Finding the defective gene. Nature 1986;321:558–559.
17   Capron M: Eosinophils; receptors and mediators in hypersensitivity. Clin Exp Allergy 1989;19(suppl 1):3–8.
18   Corrigan CJ, Kay AB: T cells and eosinophils in the pathogenesis of asthma. Immunol Today 1992; 13:501–507.
19   Coffman RL, Ohara J, Bond MW: B cell stimulatory factor-I enhances IgE response of lipopoly-saccharide-activated B cells. J Immunol 1986;136:4538–4541.
20   NIH CEPH Collaborative Mapping Group: A comprehensive genetic linkage map of the human genome. Science 1992;258:67–86.
21   Matise TC, Perlin M, Chakravarti A: Automated construction of genetic linkage maps using an expert system (MultiMap): A human genome linkage map. Nat Genet 1994;6:384–390.
22   Reed PW, Davies JL, Copeman JB, Bennett ST, Palmer SM, Pritchard LE, Gough SC, Kawaguchi Y, Cordell HJ, Balfour KM: Chromosome-specific microsatellite sets for fluorescence-based, semi-automated genome mapping. Nat Genet 1994;7:390–395.

23 Weissenbach J: A second-generation linkage map of the human genome based on highly informative microsatellite loci. Gene 1993;135:275–278.

24 Riordan JR, Rommens JM, Kerem B-S, Alon N: Identification of the cystic fibrosis gene: Cloning and characterisation of complementary DNA. Science 1989;245:1066–1073.

25 Rommens JM, Iannuzzi MC, Kerem B-S, Drumm MJ: Identification of the cystic fibrosis gene: Chromosome walking and jumping. Science 1989;245:1059–1065.

26 Welsh MJ, Smith AE: Molecular mechanisms of CFTR chloride channel dysfunction in cystic fibrosis. Cell 1993;73:1251–1254.

27 Hyde SC, Gill DR, Higgins CF, Trezise AEO, MacVinish LJ, Cuthbert AW, Radcliff R, Evans MJ, Colledge WH: Correction of the ion transport defect in cystic fibrosis transgenic mice by gene therapy. Nature 1993;362:250–255.

28 Morton NE: Major gene for atopy? Clin Exp Allergy 1992;22:1041–1043.

29 Haseman JK, Elston RC: SAGE (Statistical Analysis for Genetic Epidemiology). Behav Genet 1972; 2:3.

30 Blackweider WC, Elston RC: A comparison of sib-pair linkage tests for disease susceptibility loci. Genet Epidemiol 1985;2:85–97.

31 Cookson WOCM, Sharp PA, Faux JA, Hopkin JM: Linkage between immunoglobulin E responses underlying asthma and rhinitis and chromosome 11q. Lancet 1989,i:1292–1295.

32 Young RP, Sharp PA, Lynch JR, Faux JA, Lathrop GM, Cookson WOCM, Hopkin JM: Confirmation of genetic linkage between atopic IgE responses and chromosome 11q13. J Med Genet 1992; 29:236–238.

33 Cookson WOCM, Young RP, Sandford AJ, Moffatt MF, Shirakawa T, Sharp PA, Faux JA, Julier C, Le Souef PN, Nakamura Y, Lathrop GM, Hopkin JM: Maternal inheritance of atopic IgE responsiveness on chromosome 11q. Lancet 1992;340:381–384.

34 Sandford AJ, Shirakawa T, Moffatt MF, Daniels SE, Ra C, Faux JA,Young RP, Nakamura N, Lathrop GM, Cookson WOCM, Hopkin JM: Localization of atopy and β-subunit of high-affinity IgE receptor (FcεR1) on chromosome 11q. Lancet 1993;341:332–334.

35 Moffatt MF, Sharp PA, Faux JA, Young RP, Cookson WOC, Hopkin JM: Factors confounding genetic linkage between atopy and chromosome 11q. Clin Exp Allergy 1992;22:1046–1051.

36 Magnusson CG: Cord serum IgE in relation to family history and as predictor of atopy disease in early infancy. Allergy 1988;43:241–251.

37 Arshad SH, Matthews S, Gant C, Hide DW: Effect of allergen avoidance on development of allergic disorders in infancy. Lancet 1992;339:1493–1497.

38 Halonen M, Stern D, Taussig LM, Wright A, Ray CG, Martinez FD: The predictive relationship between serum IgE levels at birth and subsequent incidences of lower respiratory illnesses and eczema in infants. Am Rev Respir Dis 1992;46:866–870.

39 Lympany P, Welsh KI, Cochrane GM, Kemeny DM, Lee TH: Genetic analysis of the linkage between chromosome 11q and atopy. Clin Exp Allergy 1992;22:1085–1092.

40 Hizawa N, Ohe M, Yamaguchi E, Itoh A, Furuya K, Ohnuma N, Munakata M, Kawakami Y: Lack of linkage between atopy and locus 11q13. Clin Exp Allergy 1992;22:1065–1069

41 Rich SS, Roitman-Johnson B, Greenberg B, Roberts S, Blumenthal MN: Genetic analysis of atopy in three large kindreds: No evidence of linkage to D11S97. Clin Exp Allergy 1992;22:1070–1076.

42 Amelung PJ, Panhuysen CIM, Postma DS, Levitt RC, Koeter GH, Francomano CA, Bleeker ER, Meyers DA: Atopy and bronchial hyperresponsiveness: Exclusion of linkage to markers on chromosomes 11 and 6p. Clin Exp Allergy 1992;22:1077–1084.

43 Shirakawa T, Hashimoto T, Furuyama J, Morimoto K: Linkage between severe atopy and chromosome 11q13 in Japanese families. Clin Genet 1994;46:228–232.

44 Collee JM, de Vries HG, Gerritsen J: Allele sharing on chromosome 11q13 in sibs with asthma. Lancet 1993;ii:936.

45 Tedder TF, Streuli M, Schlossman SF, Saito, H: Isolation and structure of cDNA encoding the B1 (CD20) cell-surface antigen of human B lymphocytes. Proc Natl Acad Sci USA 1988;85:208–212.

46 Charmley P, Nguyen J, Tedder TF, Gatti R: A frequent human CD20 (B1) differentiation antigen DNA polymorphism detected with MspI is located near 11q12–13. Nucleic Acids 1990;18:207.

47   Blank U, Ra C, Miller L, White K, Metzger H, Kinet JP: Complete structure and expression in tranfected cells of high affinity IgE receptor. Nature 1989;337:187–189.

48   Ra C, Jouvin MHE, Kinet JP: Complete structure of the mouse mast cell receptor for IgE (FcεRI) and surface expression of chimeric receptors (rat-mouse-human) on transfected cells. J Biol Chem 1989;264:15323–15327.

49   Bieber T, de la Salle H, Wollenberg A, Hakimi J, Chizzonite R, Ring J, Hanau D, de la Salle C: human epidernal Langerhans' cells express the high-affinity receptor for immunoglobulin E (FcεRI). J Exp Med 1992;175:1285–1290.

50   Maurer D, Fiebiger E, Reininger B, Wolff-Winiski B, Jouvin MH, Kolgus O, Kinet J-P, Stingl G: Expression of functional high-affinity immunoglobulin E receptors (FcεRI) on monocytes of atopic individuals. J Exp Med 1994;179:745–750.

51   Oliver JM, Seagrave J, Stump RF, Pfeiffer JR, Deanin GG: Signal transduction and cellular response in RBL-2H3 mast cells. Prog Allergy 1988;42:185–245.

52   Helm B, Marsh P, Vercelli D, Padlan E, Gould G, Geha R: The mast cell binding site on human immunoglobulin E. Nature 1988;331:180–183.

53   Plaut M, Pierce JH, Watson CJ, Hanley-Hyde J, Nordan RP, Paul WE: Mast cell lines produce lymphokines in response to cross-linkage of (FcεRI) to calcium ionophores. Nature 1989;339:64–67.

54   Ben-Sasson SZ, Le Gros G, Conrad DH, Finkelman FD: Cross-linking Fc receptors stimulate splenic non-B, non-T cells to secrete interleukin-4 and other lymphokines. Proc Natl Acad Sci USA 1990;87:1421–1425.

55   Thompson HL, Metcalfe DD, Kinet JP: Early expression of high-affinity receptor for immunoglobulin E (FcεRI) during differentiation of mouse mast cells and human basophils. J Clin Invest 1990; 85:1227–1233.

56   Wodnar-Filipowicz A, Heusser CH, Moroni C: Production of the haemopoietic growth factors GM-CSF and interleukin-3 by mast cells in response to IgE receptor-mediated activation. Nature 1989;339:150–152.

57   Paul WE, Seder RA, Plant M: Lymphokine and cytokine production by FcεRI + cells. Adv Immunol 1992;53:1–29.

58   Gascan H, Gauchat JF, Roncarolo MG, Yssel H, Spits H, de Vries JE: Human B cell clones can be induced to proliferate and to switch to IgE and IgG4 synthesis by interleukin -4 and a signal provided by activated CD4 + T cell clones. J Exp Med 1991;17:747–750.

59   Gauchat J-F, Henchoz S, Mazzel G, Aubry J-P, Brunner T, Blasey H, Life P, Talabot D, Flores-Romo L, Thompson J, Kishi K, Butterfield J, Dahinden C, Bonnefoy J-Y: Induction of human IgE synthesis in B cells by mast cells and basophils. Nature 1993;365:340–343.

60   Shirakawa T, Li A, Dubowitz M, Dekker JW, Shaw AE, Faux JA, Ra C, Cookson WOCM, Hopkin JM: Association between atopy and variants of the β-subunit of the high-affinity immunoglobulin E receptor. Nat Genet 1994;7:125–130.

61   Ferrie RM, Schwarz MJ, Robertson NG, Vaudin S, Super M, Malone G, Little S: Development, multiplexing, and application of ARMS tests for common mutations in the CFTR gene. Am J Mum Genet 1992;51:251–262.

62   Kinet J-P, Quarto R, Perez-Montfort R, Metzger H: Non-covalently and covalently bound lipid on the receptor for immunoglobulin E. Biochemistry 1985;24:7342–7348.

63   Dohlman HG, Caron MG,, Lefkowitz R: A family of receptors coupled to guanine nucleotide regulatory proteins. Biochemistry 1987;26:2657–2663.

64   Rees DC, DeAntonio L, Eisenberg D: Hydrophobic organizaiton of membrane proteins. Science 1989;245:510–513.

65   Varin-Blank N, Metzger H: Surface expression of mutated subunits of the high-affinity mast cell receptor for IgE. J Biol Chem 1990;265:15685–15694.

66   Oksenberg D, Marsters SA, O'Dowd BF, Jin H, Havlik S, Peroutka SJ, Ashkenazi A: A single amino-acid difference confers major pharmacological variation between human and rodent 5-HT$_{1B}$ receptors. Nature 1992;360:161–163.

67   Shenker A, Laue L, Kosugi S, Merendino JJ Jr, Minegishi T, Cutler GB Jr: A constitutively activating mutation of the leutenizing hormone receptor in familial male precocious puberty. Nature 1993; 365:652–654.

68   Beinborn M, Lee Y-M, McBride EW, Quinn SM, Kopin AS: A single amino acid of the cholecystokinin-B/gastrin receptor determines specificity for non-peptide antagonists. Nature 1993;362:348–350.
69   Eccleston E, Leonard BJ, Loee JS, Wellford HJ: Basophilic leukaemia in the albino rat and a demonstration of basoprotein. Nature New Biol 1973;244:73–76.
70   Kinet JP, Blank U, Ra C, White K, Metzger H, Kochan J: Isolation and characterisation of cDNAs coding for the beta subunit of the high-affinity receptor for immunoglobulin E. Proc Natl Acad Sci USA 1988;85:6483–6487.
71   Moffatt MF, Hill MR, Cornelis F, Schou C, Faux JA, Young RP, James AL, Tyan G, le Souef P, Musk AW, Cookson WOCM, Hopkin JM: Genetic linkage of the TCR-α/δ region to specific immunogulobulin E responses. Lancet 1994;343:1597–1600.
72   Marsh DG, Neely JD, Breazeale DR, Ghosh B, Friedhoff LR, Ehrlich-Kautzky E, Schou C, Krishmaswamy G, Beaty TG: Linkage analysis of IL-4 and other chromosome 5q31.1 markers and total serum immunoglobulin E concentrations. Science 1994;264:1152–1155.

J.M. Hopkin, MD, FRCP, Consultant Physician, Osler Chest Unit, Churchill Hospital, Oxford OX3 7LJ (UK)

Hall, IP (ed): Genetics of Asthma and Atopy.
Monogr Allergy. Basel, Karger, 1996, vol 33, pp 109–124

....

Candidate Loci and Random Marker Approaches to Studying the Genetics of Asthma

Jane Wilkinson, Stephen T, Holgate

University Medicine, Southampton General Hospital, Southampton, UK

Introduction

The study of the genetics of atopy and asthma is hampered by the lack of consistency in research methodologies and the lack of clear agreement on definition of the phenotype. Atopy is defined as a disorder of IgE response to common allergens such as pollen, animal dander, house dust mite and fungi [1], but a patient's atopic status is designated in different ways in different studies making it difficult to confirm or refute positive findings of linkage or association to various candidate genes. Asthma can be diagnosed on the basis of clinical history, pathological changes in the bronchi, or the presence of bronchial hyperresponsiveness (BHR), but none of these is completely reliable. For example, most asthmatics give a history of wheeze, cough or shortness of breath, but these symptoms are not exclusive to asthmatics and may occur in chronic obstructive airways disease in adults and during viral upper respiratory tract infections in children. Characteristic pathological changes occur in the airways of asthmatic patients; thickening of the basement membrane, mast cell and eosinophil infiltration and desquamation of the epithelium [2], but, again, these changes are not exclusive to asthma. In addition, it would be impractical to obtain bronchial biopsies on all subjects in the large populations required for genetic studies.

BHR is frequently used as a marker of asthma. An individual is said by convention to demonstrate BHR if the forced expiratory volume in 1 s (FEV_1) falls by 20% at a given dose of inhaled histamine or methacholine, although different doses have been used by different workers. On the basis of this, one would hope to be able to discriminate clearly between asthmatics and nonasthmatics. However, some atopic subjects with no evidence of symptomatic asthma will also demonstrate BHR, as will a small percentage of normal subjects. Asthmatics with definite symptoms of asthma and clear evidence of airflow obstruction may not demonstrate BHR, and, conversely, asthmatic subjects who are symptom-free at time of studying may exhibit greatly enhanced BHR [3–6].

Relationshsip between Atopy, BHR and Asthma

Atopy is a major risk factor for the development of asthma, eczema and allergic rhinitis, and is detected by a raised total serum IgE, a raised specific IgE and positive skin tests to common aeroallergens. The relationship between IgE levels and asthma was examined in a study by Burrows et al. [7], and a close association was found between self-reported asthma and serum total IgE standardized for age and sex. No asthma was present in the individuals with the lowest IgE levels for their age and sex. Sears et al. [8] confirmed this relationship in a study of IgE levels and BHR to methacholine challenge and asthma in a birth cohort of New Zealand children. The prevalence of diagnosed asthma was significantly related to the serum IgE level. Furthermore, in children who had been asymptomatic throughout their lives and who had no history of atopic disease, BHR was still found to be related to an allergic diathesis as reflected by the serum total IgE level. Further work by Sears et al. [9, 10] on the same cohort of children has examined the relationship between BHR, atopy and asthma. BHR was found to be strongly correlated with reported asthma and wheezing and atopy as defined by positive skin prick tests, especially to cat and house dust mite. It is clear from this work that atopy, BHR and asthma are closely related, but the precise nature of the link remains unclear.

A number of variables have been shown to affect both serum total IgE levels and BHR. Smoking, for example, has been shown to lead to an elevation in total serum IgE levels [11]. The effect of age on serum IgE levels is unclear with studies showing variously, a decline, an increase or no change [12]. Environmental factors undoubtedly influence basal serum levels of IgE and will vary depending on the time of year the sample was taken. Ideally, therefore, multiple samples should be taken and looked at for seasonal variation. BHR may also be affected by smoking and previous history of respiratory illness [13, 14]. These variables should be taken into account in any analysis which uses serum total IgE or BHR as surrogate markers of atopy and asthma.

Both serum total IgE and BHR are continuous variables, and the cut-off point between normal and abnormal is, in some senses, arbitrary. Cookson and Hopkin [15] use a broad definition of atopy in their work; an individual is designated atopic it they are found to have one of the following: one positive skin prick test, a serum IgE level >2.5 SD above the mean for the normal population or one or more positive radioallergsorbent tests (RASTs). Shirakawa et al. [16] used a more stringent definition of atopy that required all three criteria to be met for the individual to be designated atopic, but these definitions are subjective and force a dichotomy on a quantitative trait.

Genetics of IgE, Atopy and BHR

Despite the problems of using serum IgE and BHR in addition to clinical history as markers for the asthma phenotype, they do at least provide objective measurements which can be used as a basis for a study of the genetics of allergic disease. There is clearly a genetic component to both specific and total IgE production, but conflicting reports as to the mode of transmission have emerged. Studies looking at IgE levels in twins have found that monozygotic twins are substantially more similar to each other than are otherwise comparable dizygotic twins of a pair [17–20], and various models including single dominant gene with partial penetrance, single recessive gene with partial penetrance and multigene inheritance have all been described [21–24].

The genetics of BHR are also unresolved. A study by Townley et al. [25] has demonstrated both an environmental and a genetic component to BHR, although it was thought unlikely to be determined at a single locus.

Methods for Dissecting the Genetic Basis of Complex Traits

Linkage Analysis

Methods for dissecting the genetic basis of complex disease include linkage analysis, allele-sharing methods and association studies [26]. The first of these, linkage analysis, involves proposing a model to explain the inheritance of the phenotypes and genotypes involved in the pedigree. It can be very powerful if the correct model is applied, but if an incorrect model is applied then the results, be they positive or negative, will be difficult to interpret. Segregation analysis which involves fitting a general model to the inheritance pattern of a trait is extremely sensitive to ascertainment bias. For example, the study by Cookson and Hopkin [15] found evidence for the vertical transmission of atopy based on a sample of 20 nuclear families recruited via asthmatic probands. In addition, three large families with asthmatic members were recruited: two by means of letter to general practitioners and one in response to a media appeal.

It was felt that the inheritance pattern of atopy with particular regard to the extended families clearly indicated autosomal dominant transmission. Given the broad definition of atopy and the inclusion of extended pedigrees with such a strong history of atopy, it is difficult to reject the autosomal dominant hypothesis, but this does not necessarily imply that it is correct.

Meyers et al. [27] attempted to resolve the confusion over the genetic basis for IgE production by studying 42 families deliberately *not* selected for the presence of atopy. Segregation analysis showed that the mixed model of recessive inheritance of high levels of IgE was most appropriate for their data, with approximately 36% of the total phenotypic variation in IgE attributable to genetic factors, equally divided between Mendelian and more general polygenic components.

Allele-Sharing Methods

Allele-sharing methods are an attempt to prove that the inheritance pattern of a chromosomal region is not consistent with random Mendelian segregation by showing that affected relatives inherit identical copies of the region more often than would be expected by chance. Because allele-sharing methods are nonparametric, i.e. they assume no model for the inheritance of the trait, they tend to be more robust than linkage analysis: affected relatives should show excess allele sharing even in the presence of incomplete penetrance, phenocopy, genetic heterogeneity and high-frequency disease alleles. The trade-off is that allele-sharing methods are often less powerful than a correctly applied linkage model.

Allele-sharing methods involve studying affected relatives in a pedigree to see how often a particular copy of a chromosomal region is shared identical by descent (IBD), that is inherited from a common ancestor within a pedigree. Affected sib-pair analysis is the simplest form of this method. For example, 2 sibs can share zero, one or two copies of any locus (with 25–50–25% distribution expected under random segregation). It both parents are available, the data can be partitioned into separate IBD sharing for the maternal and paternal chromosome (zero or one copy with a 50–50% distribution expected under random segregation). In either case, excess allele sharing can be measured with a simple χ^2 test.

Allele-sharing methods can be applied to quantitative traits. An approach proposed by Haseman and Elston [28] is based on the notion that the phenotypic similarity between two relatives should be correlated with the number of alleles shared at a trait-causing locus. This method was used by Marsh et al. [29] to relate serum IgE levels with allele sharing in the region of the gene encoding interleukin (IL)-4 on chromosome 5q.

Association Studies

Association studies do not concern familial inheritance patterns at all. Rather, they are case-control studies based on a comparison of unrelated affected and unaffected individuals from a population. An allele A at a gene of interest is said to be associated with the trait if it occurs at a significantly higher frequency among affected compared with control individuals. Although association studies can be performed for any random DNA polymorphisms, they are most meaningful when applied to functionally significant variation in genes having a clear biological relation to the trait. Positive association can occur for three main reasons: it may occur if the allele is actually the cause of the disease; it can occur if the allele is in linkage disequilibrium with the disease-causing gene, and it can also occur as a result of population admixture. In a mixed population, any trait present at a higher frequency in an ethnic group will show positive association with any allele that also happens to be more common in that group. To prevent spurious associations from arising for this reason, association studies should be performed within relatively homogeneous populations.

Candidate Loci versus Random Markers

In a complex condition such as asthma, there is unlikely to be a simple correspondence between genotype and phenotype. For example, the same genotype may result in different phenotypes or, conversely, different genotypes can result in the same phenotype. In addition, the genotype at any given locus may affect the probability of developing the disease but not fully determine its outcome. Although an individual may inherit a predisposition to develop the disease, environmental factors are crucial for its expression [30].

It is almost certainly the case that asthma and atopy are genetically heterogeneous, that is a mutation in any one of several genes may result in identical phenotypes. This may hamper genetic mapping as a chromosomal region may cosegregate in some families but not in others. To complicate the matter further, some traits, such as the serum IgE and BHR, may require the presence of mutations in multiple genes for their expression, a fact which is borne out by previous attempts at segregation analysis of these traits.

Progress in this difficult field has been facilitated by the development of increasing numbers of DNA polymorphisms, which have made the mapping of disease loci very successful, even when indistinguishable phenotypes were produced by different genes in different families. Broadly speaking there are two approaches which can be used. The first of these is to screen the entire genome using highly polymorphic DNA markers positioned at intervals on each chromosome. This approach has been used successfully by Todd and co-workers [31] in their work on insulin-dependent diabetes mellitus (IDDM) using a linkage map of 290 marker loci. The fluorescence-based genome linkage map had an average spacing of 11-cM, and the marker loci had a mean heterozygosity of 0.8. Gaps exist in the map, particularly at the proterminal regions, but only about 3% of the genome remains to be covered at a 20-cM resolution. However, it was noted by Todd's group that a marker only 3-cM away from the IDDM-2 locus would not have shown linkage. therefore, they concluded that to identify all the major genes for type 1 diabetes by linkage analysis, a map with markers at 3-cM intervals would be required. Positive findings of linkage can be investigated further by looking for candidate genes in the area of interest. Negative findings, however, need to be interpreted with caution, as markers may not lie close enough to putative genes for positive linkage to be found. The Oxford group found linkage to 11q13 for atopy following a search with a series of random markers, and they have since focused on the β subunit of the high-affinity Fcε receptor (FcεR1) [32, 33].

The second approach relies on focusing on candidate loci such as the cytokine cluster on chromosome 5q31.1-q33. Several key cytokines in the pathogenesis of asthma including IL-4, IL-5, granulocyte-macrophage colony-stimulating factor (GM-CSF) and IL-9 are encoded in this cluster. Marsh et al. [29] looked for linkage between total serum IgE and multiallergen IgE antibodies and several polymorphic genetic markers in and around chromosome 5q31.1-q33, with a primary focus on markers mapping within the IL-4 gene itself and in or close to the IL-4 cluster. Using the sib-pair method of analysing the data, Marsh et al. found significant evidence for linkage for IL-4, interferon-releasing factor 1 (IRF-1), IL-9, and two anonymous markers in the same region, D5S393 and D5S399 with serum total IgE.

Both the findings on 11q and 5q have been followed by a series of negative and positive reports which highlight the need for international collaborative efforts in this perplexing field.

Genetics of Asthma in a Wessex Population

We have tried to address some of the problems outlined above in our study of the genetics of allergic disease in Wessex [34, 35]. We have collected two separate samples of families, the first recruited at random from the population and the second recruited on the basis of two or more members affected by asthma. It was felt necessary to recruit a random sample, as both atopy and asthma are common conditions; the prevalence of asthma in men has increased from 11.6 to 20.5 per 1,000 population from 1970 to 1981 and in women from 8.8 to 15.9 per 1,000 population. The prevalence of hay fever has also increased from 10.8 to 19.8 per 1,000 population in men and 10.3 to 19.7 in women over the same period [36]. Genes which are thought to be important in the pathogenesis of allergic disease should also be found in a population recruited at random with no reference to the disease or trait in question. Positive findings in a population enriched for the disease phenotype need to be replicated in a random sample for an estimate of their probable true importance.

On account of this we recruited a random cohort of 131 families with 3 or more children ascertained through general practitioner registers in the Southampton district. There was no selection for asthma or atopy, and the families were recruited to take part in a health survey without specifying its focus. The clinical status of each family member in both samples with respect to asthma and atopy was recorded on a structured questionnaire, and laboratory indices of bronchial responsiveness and atopy were determined. The questionnaire was based on the IUATLD standard for adults and the ISAAC protocol for children [37]. In addition, to ensure a clear understanding by the subjects, a questionnaire was also completed with the aid of a video [38]. Total IgE in the serum was determined by a standard RAST. Specific IgE status was determined by skin prick testing to 11 common allergens with a negative (saline) and positive (histamine) control. The major and minor axes of the wheal were recorded. The allergens used were *Dermatophagoides pteronyssinus,* cat, dog, horse, mixed grass pollen, tree pollen, *Aspergillus fumigatus, Alternaria, Cladasporium,* milk and egg (white and yolk). The bronchial response to inhaled histamine acid phosphate was measured using the hand-held nebulizer procedure modified from that of Yan et al. [39]. After an initial saline inhalation, doubling concentrations of agonist were administered until FEV_1 fell by 20% of the post-saline baseline or the highest concentration of agonist (2.2 μmol) was reached. FEV_1 was recorded using a Vitalograph spirometer 5 min after completing each aerosol inhalation, the best of three consecutive readings being recorded. Histamine responsiveness was quantified both as the provocation concentration of agonist that reduced FEV_1 by 20% of the starting base (PC_{20}) and the AUC of the dose-response curve.

Initially we chose to type markers in and around the 11q locus in an attempt to replicate the Oxford group's findings of linkage to 11q with atopy. Three highly informative polymorphic markers, D11S534, D11S527, and D11S480, were typed in the sample of random families. In addition, markers were chosen for their proximity to candidate loci on chromosomes 1, 5, 6, 9, 12, 16, 19 and 22 (table 1).

DNA was extracted from peripheral blood leucocytes, and multiplex polymerase chain reaction (PCR) was initially performed using primers to markers

Table 1. Markers typed in the Wessex population with chromosomal location and closest candidate loci

Locus	Name	Location	Closest candidate	Number of alleles
D11S480		11q13.1	$FC_\varepsilon R1\beta = CD2$	9
D11S527		11q13.5	$DC_\varepsilon R1\beta = CD2$	13
D11S534		11q13	$FC_\varepsilon R1\beta\text{-}CD2$	14
D16S298		16p12.1–11.2	IL-4R	11
C19S112		19q13.3–13.4	IL-11	11
D19S177		19p13.3	$CD23 = FC_\varepsilon R2$	10
D1S104		1q21–23	$FC_\varepsilon R1\alpha,\gamma$	9
IFN-α	Interferon-α	9p22	IFN-α	6
IGF-1	Insulin-like growth factor	12q22–24	IFN-γ	7
IL-2R-β	Interleukin-2 receptor-β	22q13	IL-2R-β = CD25	12
IL-9	Interleukin-9	5q23–31	IL-9, IL-4	11
TNF-β	Tumour necrosis factor-β	6p21.3	TNF-β	14

in the vicinity of candidate loci listed in table 1. One primer from each pair was labelled with one of three fluorescent dye esters: Tamra, Joe and Fam. The products were run on a urea 6% polyacrylamide gel in a model 373A DNA sequencer using version 672 Genescan software (Applied Biosystems Inc., Warrington, UK). Each marker locus produced a discrete range of labelled products; those in the same dye colour being easily distinguished. Internal size standards allowed the alleles to be sized accurately and reproducibly to within one base pair.

We have previously used stepwise principal component regression analysis of six traits (log serum total IgE, skin prick test, BHR, history of wheeze, history of eczema and history of seasonal rhinitis) to define atopy [40] (table 2). Atopy was defined as the derived first principal component of the age- and sex-adjusted traits and was dominated by serum total IgE. Asthma was defined in a similar manner using stepwise principal component regression of BHR

Table 2. Trait correlations

Trait	IgE	BR	SP	WZ	EZ	HF	MG	AY
Log IgE (IgE)	1.00	0.39	0.54	0.39	0.28	0.30	0.13	0.98
Bronchial reactivity (BR)		1.00	0.38	0.63	0.14	0.27	0.09	0.49
Skin prick (SP)			1.00	0.45	0.19	0.43	0.07	0.67
Wheeze (WZ)				1.00	0.23	0.34	0.03	0.49
Eczema (EZ)					1.00	0.13	0.09	0.30
Hay fever (HF)						1.00	–0.09	0.38
Migraine (MG)							1.00	0.13
Atopy (AY)								1.00

and history of wheeze and was dominated by BHR. Our analysis is therefore based on BHR to histamine and the logarithm of serum total IgE corrected for age and sex.

Nonparametric linkage and association analysis was performed with the NOPAR programme that calculated a z score for each locus and allele. Parametric analyses were performed with the COMDS programme on an ordered polychotomy of ranked scores, corresponding to percentiles 31, 56, 72, 82, 89, 94, 97, 99, 100 from fathers, mothers and children separately. 131 families agreed to be studied, comprising 685 individuals (262 parents and 423 offspring). Mean age of the parents was 41.5 years (range 31–58) and of the children 12.9 years (range 2–30). Of the parents, 17 (6.5%) had self-reported asthma, while 83 (19.8%) of children had self- or parent-reported asthma, consistent with asthma prevalence in the Southampton area. 107 (40.8%) of the parents and 189 (45.7%) of the children had a skin prick test of mean wheal diameter of 3 mm or greater. The geometric mean serum total IgE was 276 IU (range 5–2,800) for the parents and 300 IU (range 2–10,400) for the children. Two parents and 4 children had a baseline FEV_1 below 60% and so were unable to perform the histamine challenge. Twenty-seven (10.4%) of the parents and 70 (16.7%) of the children demonstrated a PD_{20} to the highest cumulative dose of histamine.

In our analysis of the three markers on 11q, we found no evidence of linkage for atopy using both parametric and nonparametric tests [40]. However, the markers are distal to the $Fc_\varepsilon R1$ locus, and this might explain why we found no evidence to support linkage. Allelic association tests were, however, positive for two of the loci on 11q. At the D11S527 locus, allele 168 was found to be significantly associated with bronchial hyperreactivity (p=0.0003) but showed no association with atopy, while at the D11S534 locus, allele 235 was significantly associated with the serum total IgE (p = 0.007) but showed no association with BHR. At the IL-9 locus on chromosome 5q, the 118 allele showed significant association with serum total IgE (p < 0.003) but not with BHR. Given the number of markers tested, in some cases significant association is to be expected by chance and may be simply due to a type 1 error. For the association with IL-9, there are two contrary arguments. Firstly the interleukin cluster around IL-9 contains many of the most attractive candidate loci, including IL-4. The candidate loci in this region have biological plausibility, and, secondly, linkage for total serum IgE has previously been reported in this region of chromosome 5q in two quite different populations [29, 41]. The study by Marsh et al. [29] was carried out on a small sample of 11 Caucasian Amish families, a genetically isolated farming community. The study by Meyers et al. [41] was based on a sample recruited for asthma and studied for atopy. Our study has given further evidence for the potential importance of this region of the human genome in a large sample recruited at random from the population.

Despite the poor characterization of the physical map of this region of 5q31, there remain a number of strong candidate loci near the reported markers (fig. 1). The organization of these genes on the chromosome is well conserved between species [42], suggesting an important role in the regulation of their expression. Many of the candidate loci share a common intron-exon structure, most notably IL-4, IL-5, GM-CSF [43] and IL-13 [42]. IL-4 has the greatest biological plausibility given its critical role in isotype switching of B cells to IgE production [44], its capacity to determine the maturation of CD4 + T cells along the Th2 pathway [45–47], and to up-regulate selectively the expression of the vascular adhesion molecule 1 (VCAM-1), which, through an interaction with the integrin VLA-4 ($\alpha_4\beta_1$) is responsible for the selective recruitment and priming of eosinophils in allergic inflammatory responses [48]. Other candidates warrant consideration. IL-13 has a 20% sequence homology with IL-4 and shares many of its biological activities including B-cell immunoglobulin class switching to IgE [49]. It appears to lie upstream of IL-4 within 50 kb [50]. IL-5, which promotes eosinophil maturation and priming, is 110–180 kb downstream of IL-4 [50], while the closely linked IL-3 and GM-CSF genes lie a minimum of 600 kb downstream, all within close proximity to IL-9. IRF-1 is a transcription factor for interferon-inducible genes [51] and could influence the expression of the atopic phenotype through interferon-mediated down-regulation of the Th2 responses on IgE [52]. The 7-transmembrane β_2-adreno-ceptor gene lies in close proximity to the IL-4 gene cluster, and polymorphisms within it have been associated with asthma severity [53], but a direct effect on IgE responses seems unlikely. Although IL-9 may enhance the effect of IL-4-dependent IgE synthesis, it is not the most plausible candidate. The positive

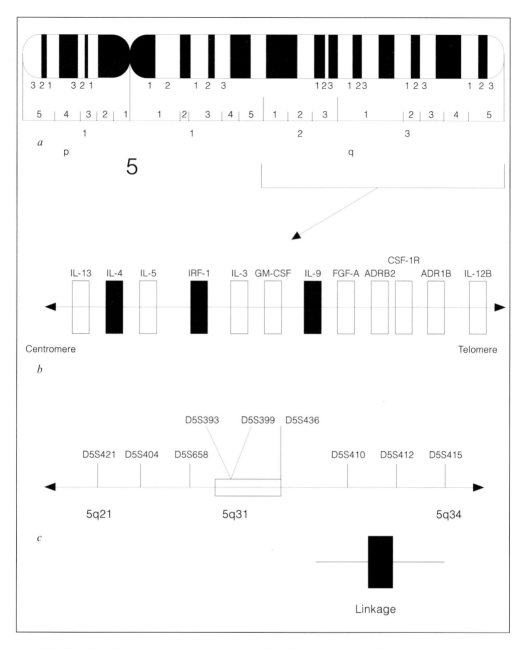

Fig. 1. a A schematic representation of the banding pattern on chromosome 5. *b* Simplified genetic map of 5q21-q34 showing IL-4 gene cluster. *c* Map of genetic markers on chromosome 5. IRF-1 = Interferon-releasing factor; GM-CSF = granulocyte macrophage-colony-stimulating factor; FGF-A = fibroblast growth factor acidic; CSF-1R = colony-stimulating factor, ADRB2 = β_2-adrenergic receptor; ADR1B = β_1-adrenergic receptor.

association may be as a result of linkage disequilibrium with a disease-causing trait, but the abundance of candidate loci in 5q31 will complicate the resolution of the true effect from linkage disequilibrium.

Atopy, transmitted exclusively through the maternal line, has been mapped to chromosome 11q13 by linkage and sib-pair analysis with the β chain of the high-affinity IgE receptor being selected as the candidate locus [33, 54, 55]. The findings were then extended to a single base pair substitution leading to a leucine residue replacing an isoleucine residue at position 181 in the fourth transmembrane domain of the β subunit of the high-affinity IgE receptor [32]. Subsequent studies have given contradictory results with some investigators reporting evidence of linkage to chromosome 111q13 [56], while a number have failed to replicate linkage [57–62], and, to date, the association of atopy with mutations within the high-affinity IgE receptor remain unconfirmed. Our finding of significant allelic association with BHR at D11S527 and IgE at D11S534 is intriguing. Although these markers are in the same region as FcεR1β, physically they are a considerable distance away. Our observation of significant allelic association in the absence of significant linkage is not unusual but reflects the greater sensitivity of the test as well as the possibilities for error at a multiallelic locus. This has also shown to the the case in the work on IDDM susceptibility genes [31]. It was found that, because of the polygenic nature of the condition being studied, evidence for linkage for even an established IDDM locus, such as IDDM-2, has not been found in every data set. However, strong evidence for association may be found for the same region in data sets that show no deviation from the expected proportion of affected sib pairs sharing 2, 1, or 0 alleles identical by descent, which is the traditional way of assessing linkage to disease in affected sib pairs.

When a positive association has been found, it should be subjected to a transmission disequilibrium test, one of the most powerful family-based association methods, detecting both linkage and association. This test has the premise that a parent heterozygous for an associated allele A_1 and a nonassociated allele A_2 should more often transmit A_1 than A_2 to an affected child [26]. This test should not be applied to the sample in which the initial association was found, as affected individuals necessarily have an excess of the associated allele, but rather to a new sample from the same population.

Association studies are not well suited to whole-genome searches in large, mixed populations. Because linkage disequilibrium extends over very short distances in an old population, one would need thousands of genetic markers to cover the genome. Moreover, testing many markers raises the serious problem of multiple hypothesis testing: each association test is nearly independent. Testing n loci each with k alleles amounts to performing about $n(k-1)$ independent tests, and the required significance level must be divided by this factor [26]. The same multiple testing issue applies to retesting results in a second sample at a later stage. An illustration of this is found in the study by Copeman et al. [63] on linkage of a type 1 diabetes susceptibility gene to chromosome 2q. Positive association was found for certain alleles and IDDM, but the significance level fell when multiple testing was taken into account. However, as the region was felt to contain possible candidate genes the association was pursued in further samples.

The Future

We are currently attempting to replicate the findings from the random population sample in the sample of families with two or more members affected by asthma. We have genotyped all the family members for markers on chromosome 5q, the cytokine cluster, in addition to the D11S527 marker on chromosome 11q which showed positive association for BHR. We have also added an additional marker from the intron of the β chain of the high-affinity IgE receptor in a further attempt to validate the Oxford work.

We are also keen to characterize the members of the multiplex families more fully by assessing bronchial responsiveness to adenosine 5′-monophosphate in addition to histamine. Adenosine is a potent bronchoconstrictor in asthmatic patients, acting predominantly by stimulating the release of mediators from mast cells [64]. It has been shown, both in adults and children, that BHR to adenosine is more specific to asthma [65] whereas histamine and methacholine challenges may be positive in patients with chronic obstructive airways disease as well as asthma.

In addition to the markers already under investigation, we intend to perform a genome wide search using the markers developed by Todd [66]. This will be supplemented wherever possible by markers within candidate genes.

The real challenge is to develop a methodology which allows us to identify and minimize the causes of divergent results in the field of complex inheritance. Methods to defect genes of small effect, which overall contribute to a disease phenotype, are as controversial now as major gene efforts were 50 years ago. Each step in the process (sampling, phenotype definition and analysis) is disputed. Poisitive findings are frequently unrepeatable, although it is often difficult to say whether this is due to real genetic difference or due to methodological differences. Until these problems are resolved, genuine progress will be slow. It is precisely this kind of undertaking that requires collaborative efforts between research groups both nationally and internationally.

References

1 Pepys J: 'Atopy': A study in definition. Allergy 1994;49:397–399.
2 Djukanovic R, Roche WR, Wilson JW, et al: Mucosal inflammation in asthma. Am Rev Respir Dis 1990;142:434–457.
3 Boulet L-P, Cartier A, Thomson NC, et al: Asthma and increases in nonallergic bronchial responsiveness from seasonal pollen exposure. J Allergy Clin Immunol 1983;71:399–406.
4 Deal EC, McFadden ER, Ingram RH, Breslin FJ, Jaeger JJ: Airway responsiveness to cold air and hyperpnea in normal subjects and in those with hay fever and asthma. Am Rev Respir Dis 1980; 121:621–628.
5 Josephs LK, Gregg I, Mullee MA, Holgate ST: Nonspecific bronchial reactivity and its relationship to the clinical expression of asthma. A longitudinal study. Am Rev Respir Dis 1989;140:350–357.
6 Juniper EF, Frith PA, Hargreave FE: Airway responsiveness to histamine and methacholine: Relationship to minimum treatment to control symptoms of asthma. Thorax 1981;36:575–579.
7 Burrows B, Fernando D, Martinez MD, Halonen M, Barbee RA, Cline MG: Association of asthma with serum IgE levels and skin test reactivity to allergens. N Engl J Med 1989;320:271–277.
8 Sears MR, Burrows B, Flannery EM, Herbison GP, Hewitt CJ, Holdaway MD: Relation between airway responsiveness and serum IgE in children with asthma and in apparently normal children. N Engl J Med 1991;325:1067–1071.
9 Sears MR, Burrows B, Herbison GP, Holdaway MD, Flannery EM: Atopy in childhood. II. Relationship to airway responsiveness, hay fever and asthma. Clin Exp Allergy 1993;23:949–956.
10 Sears MR, Burrows B, Herbison GP, Flannery EM, Holdaway MD: Atopy in childhood. III. Relationship with pulmonary function and airway responsiveness. Clin Exp Allergy 1993;23:957–963.
11 Omenaas E, Bakke P, Elsayed S, Hanoa R, Gulsvik A: Total and specific serum IgE levels in adults: Relationship to sex, age and environmental factors. Clin Exp Allergy 1994;24:530–539.
12 Villar T, Holgate ST: IgE, smoking and lung function. Clin Exp Allergy 1994;24:508–510.
13 Weiss ST, Tager IB, Munoz A, Speizer FE: The relationship of respiratory infections in early childhood to the occurrence of increased levels of bronchial responsiveness and atopy. Am Rev Respir Dis 1985;131:573–578.
14 O'Connor GT, Weiss ST, Tager IB, Speizer FE: The effect of passive smoking on pulmonary function and nonspecific bronchial responsiveness in a population-based sample of children and young adults. Am Rev Respir Dis 1987;135:800–804.
15 Cookson WOCM, Hopkin JM: Dominant inheritance of atopic immunoglobulin E responsiveness. Lancet 1988;1:86–88.
16 Shirakawa T, Hashimoto T, Furuyama J, Morimoto K: Linkage between severe atopy and chromosome 11q13 in Japanese families. Clin Genet 1994;46:228–232.
17 Bazaral M, Orgel A, Hamburger RN: Genetics of IgE and allergy: Serum IgE levels in twins. J Allergy Clin Immunol 1974;54:288–304.
18 Wuthrich B, Baumann RA, Fries RA, Schnyder UW: Total and specific IgE (RAST) in atopic twins. Clin Allergy 1981;11:147–154.
19 Townley R, Bewtra A, Watt G, Burke BS, Carney BS, Nair N: Comparison of allergen skin-test responses in monozygous and dizygous twins. J Allergy Clin Immunol 1980;65:214.
20 Hanson B, McGue M, Roitman-Johnson B, Segal NL, Bouchard TJ, Blumenthal MN: Atopic disease and immunoglobulin E in twins reared apart and together. Am J Hum Genet 1991;48:873–879.
21 Gerrard JW, Rao DC, Morton NE: A genetic study of immunoglobulin E. Am J Hum Genet 1978;30:46–58.
22 Ott J: Maximum likelihood estimation by counting methods under polygenic and mixed models in human pedigrees. Am J Hum Genet 1980;41:161–175.
23 Blumenthal MN, Namboodiri K, Mendell N, Gleich G, Elston RC, Yunis E: Genetic transmission of serum IgE levels. Am J Med Genet 1981;10:219–228.
24 Hasstedt SJ, Meyers DA, Marsh DM: Inheritance of immunoglobulin E: Genetic model fitting. Am J Med Genet 1983;14:61–66.

25   Townley RG, Bewtra A, Wilson AF, et al: Segregation analysis of bronchial response to methacholine inhalation challenge in families with and without asthma. J Allergy Clin Immunol 1986;77:101–107.

26   Lander ES, Schork NJ: Genetic dissection of complex traits. Science 1994;265:2037–2048.

27   Meyers DA, Beaty TH, Freidhoff LR, Marsh DG: Inheritance of total serum IgE (basal levels) in man. Am J Hum Genet 1987;41:51–62.

28   Haseman JK, Elston RC: The investigation of linkage between a quantitative trait and marker locus. Behave Genet 1972;2:1–19.

29   Marsh DG, Neely JD, Breazeale DR, et al: Linkage analysis of IL-4 and other chromosome 5q31.1 markers and total serum immunoglobulin E concentrations. Science 1994;264:1152–1156.

30   Newman-Taylor AJ: Environmental determinants of asthma. Lancet 1995;345:296–299.

31   Davies JL, Kawaguchi Y, Bennett ST, et al: A genome-wide search for human type 1 diabetes susceptibility genes. Nature 1994;371:130–136.

32   Shirakawa T, Li A, Dubowitz M, et al: Association between atopy and variants of the β subunit of the high-affinity immunoglobulin E receptor. Genet 1994;7:125–130.

33   Sandford AJ, Shirakawa T, Moffatt MF, et al: Localisation of atopy and β subunit of high-affinity IhE receptor ($FC_\epsilon R1$) on chromosome 11q. Lancet 1993;41:332–334.

34   Doull IJM, Lawrence S, Watson M, et al: Allelic association of markers on chromosomes 5q and 11q with atopy and bronchial hyperresponsiveness. Am J Respir Crit Care Med 1996;153:332–334.

35   Watson M, Lawrence S, Collins A, et al: Exclusion from proximal 11q of a common gene with megaphenic effect on atopy. Ann Hum Genet 1995;59:403–411.

36   Fleming DM, Crombie DL: Prevalence of asthma and hayfever in England and Wales. Br Med J 1987;294:279–283.

37   Burney PGJ, Laitinen LA, Derdrizet S, et al: Validity and repeatability of the IUATLD (1984) bronchial symptoms questionnaire: An international comparison. Eur Respir J 1989;2:940–945.

38   Shaw RA, Crane J, Pearce N, et al: Comparison of a video questionnaire with the IUATLD written questionnaire for measuring asthma prevalence. Clin Exp Allergy 1992;22:561–568.

39   Yan K, Salome C, Woolcock AJ: Rapid method for measurement of bronchial responsiveness. Thorax 1983;38:760–780.

40   Lawrence S, Beasley R, Doull I, et al: Genetic analysis of atopy and asthma as quantitative traits and ordered polychotomies. Ann Hum Genet 1994;58:359–368.

41   Meyers DA, Postma DS, Panhuysen CIM, et al: Evidence for a locus regulating total serum IgE levels mapping to chromosome 5. Genomics 1994;23:464–470.

42   McKenzie ANJ, Li X, Laergaespada DA, et al: Structural comparison and chromosomal localization of the human and mouse IL-13 genes. J Immunol 1993;150:5436–5444.

43   Van Leeuwen BH, Martinson ME, Webb GC, Young IG: Molecular organization of the cytokine gene cluster, involving the human IL-3, Il-4, Il-5, and GM-CSF genes on human chromosome 5. Blood 1989;73:1142–1148.

44   Punnonen J, Aversa G, Cocks BG, et al: Interleukin-13 induces interleukin-4-independent IgG4 and IgE synthesis and CD23 expression by human B cells. Proce Natl Acad Sci USA 1993;90: 3730–3734.

45   Mosmann TR, Coffman RL: Th1 and Th2 cells: Different patterns of lymphokine secretion lead to different functional properties. Annu Rev Immunol 1989;7:145–173.

46   Ricci M, Rossi O, Bertoni M, Matucci A: The importance of Th2-like cells in the pathogenesis of airway-allergic inflammation. Clin Exp Allergy 1993;23:360–369.

47   Swain SL, Weinberg AD, English M, Huston G: IL-4 directs the development of Th2-like helper effectors. J Immunol 1990;145:3796–3806.

48   Lobb R, Helmer ME: The pathophysiological role of α4 in vivo. J Clin Invest 1994;94;1722–1728.

49   Zurawski G, De Vries JE: Interleukin-13, an interferon-4-like cytokine that acts on monocytes and B cells, but not on T cells. Immunol Today 1994;15:19–26.

50   Morgan JG, Dolganov GM, Robbins SE, Hinton LW, Lovett M: The selective isolation of novel cDNAs encoded by the regions surrounding the human interleukin 4 and 5 genes. Nucleic Acids Res 1992;20:5173–5179.

51   Itoh S, Harada H, Nakamura Y, White R, Taniguchi T: Assignment of the human interferon regulatory factor-1 gene to chromosome 5q23-q31. Genomics 1991;10:1097–1099.

52   Maggi E, Parronchi P, Manetti R: Reciprocal regulatory role of IFNγ and IL-4 on the in vitro development of human Th1 and Th2 clones. J Immunol 1992;48:2142–2147.

53   Hall IP, Wheatley A, Wilding P, Liggett SB: Association of the Glu 27 β2-adrenoceptor polymorphism with lower airway reactivity in asthmatic subjects. Lancet 1995;345:1213–1214.

54   Cookson WOCM, Sharp PA, Faux JA, Hopkin JM: Linkage between immunoglobulin E responses underlying asthma and rhinitis and chromosome 11q. Lancet 1989;i:292–1294.

55   Young RP, Sharp PA, Lynch JR, et al: Confirmation of genetic linkage between atopic IgE responses and chromosome 11q13. J Med Genet 1992;29:236–238.

56   Collee JM, ten Kate LP, de Vries HG, Kliphuis JW, Bouman K, Scheffer H: Allele sharing on chromosome 11q13 in sibs with asthma and atopy. Lancet 1993;342:936–936.

57   Inacio F, Perichon B, Desvaux FX, David B, Petre G, Krishnamoorthy R: Genetic transmission study of grass pollen sensitivity in 17 Portuguese families. J Allergy Clin Immunol 1991;87:204.

58   Amelung PJ, Panhuysen CIM, Postma DS, et al: Atopy and bronchial hyperresponsiveness: Exclusion of linkage to markers on chromosomes 11q and 6p. Clin Exp Allergy 1992;22:1077–1084.

59   Hizawa N, Yamaguchi E, Ohe M, et al: Lack of linkage between atopy and locus 11q13. Clin Exp Allergy 1992;22:1055–1059.

60   Rich SS, Roitman-Johnson B, Greenberg B, Roberts S, Blumenthal MN: Genetic analysis of atopy in three large kindreds: No evidence of linkage to D11S97. Clin Exp Allergy 1992;22:1070–1076.

61   Lympany P, Welsh KI, Cochrane GM, Kemeny DM, Lee TH: Genetic analysis of the linkage between chromosome 11q and atopy. Clin Exp Allergy 1992;22:1085–1092.

62   Coleman R, Trembath RC, Harper JI: Chromosome 11q13 and atopy and underlying atopic eczema. Lancet 1993;341:1121–1122.

63   Copeman JB, Cucca F, Hearne CM, et al: Linkage disequilibrium mapping of a type 1 diabetes susceptibility gene (IDDM-7) to chromosome 2q31-q33. Nat Genet 1995;9:80–85.

64   Holgate ST, Mann JS, Church MK, Cushley MJ: Mechanisms and significance of adenosine-induced bronchoconstriction in asthma. Allergy 1987;42:481–484.

65   Avital A, Springer C, Bar-Yishay E, Godfrey S: Adenosine, methacholine, and exercise challenges in children with asthma or paediatric chronic obstructive pulmonary disease. Thorax 1995;50: 511–516.

66   Reed PW, Davies JL, Copeman JB, et al: Chromosome-specific microsatellite sets for fluorescence-based, semi-automated genome mapping. Nat Genet 1994;7:390–395.

Dr. Jane Wilkinson, University Medicine, Southampton General Hospital,
Tremona Road, Southampton SO16 6YD (UK)

Hall, IP (ed): Genetics of Asthma and Atopy.
Monogr Allergy. Basel, Karger, 1996, vol 33, pp 125–137

Polymorphism at the Tumour Necrosis Factor Locus and Asthma

D.A. Campbell [a], *E. Li Kam Wa* [a], *J. Britton* [b], *S.T. Holgate* [c],
A.F. Markham [a], *J.F.J. Morrison* [a]

[a] Molecular Medicine Unit, Clinical Sciences Building, St. James' University Hospital,
Leeds,
[b] Department of Respiratory Medicine, City Hospital, Nottingham, and
[c] Department of Medicine, Southampton General Hospital, Southampton, UK

Introduction

Although initially described for its cytotoxic and antitumour activities, the family of related cytokines termed the tumour necrosis factors (TNFs) have now been shown to have a diverse role in the modulation of the inflammatory immune response. Although a substance showing cytotoxic and antitumour activities was initially described in the 19th century by Coley [1891] it was not until almost a century later that a group working in the same institute described what is now known as TNF-α [Carswell et al., 1975].

Carswell et al. [1975] originally described a soluble factor which could be identified in the sera of animals treated with reticuloendothelial stimulators, bacterial endotoxins or lipopolysaccharide (LPS). Mannel et al. [1981] subsequently described the macrophage as being the principal source of TNF and that this same molecule was an important effector molecule in the nonspecific tumouricidal activity of macrophages. TNF appeared to be a preferential mediator of cytoxicity for tumour and transformed cells and received its name due to its ability to cause haemorrhagic necrosis of subcutaneous tumours in mice. Thence came the expectation that TNF could provide a major key to therapy of malignant disease.

The cloning of the two original forms of TNF (TNF-α and TNF-β– now renamed TNF and lymphotoxin (LT)-α respectively) in 1984 by Gray et al. and Pennica et al. and the subsequent expression of these genes in *Escherichia coli,* allowed further work to be performed. Sugarman et al. [1985] showed that TNF was capable of stimulating proliferation of normal cells; Dayler et al. [1985] showed that TNF could stimulate collagenase production, whilst Bertolini et al. [1986] showed that TNF could cause bone resorption. A review by Aiyer and Aggarwal [1987] describes a number of other activities of TNF on normal cells including effects on the metabolism of fibroblasts, neutrophils, hepatocytes and leukocytes. Tracey et al. [1986] first suggested that TNF-α had the ability to induce similar effects to those resulting from endotoxin-induced shock, and Beutler et al. [1985] claim that TNF is the primary mediator in endotoxin-induced shock. However, in the experiments of Beutler et al. [1985], the authors fail to show a complete block of the shock following administration of anti-TNF antibodies, suggesting that other mediators are also important. TNF is now known to have a large spectrum of effects on a large number of cell types. TNF causes the adhesion of neutrophils to endothelial cells [Gamble et al., 1985] and may be the primary cause of endotoxin-induced neutropenia. TNF has also been shown to be growth inhibitory in that it causes vascular endothelial cell remodeling in which the cells flatten, become overlapping, rearrange their actin filaments and lose stainable fibronectin. In addition to causing adhesion of neutrophils to endothelial cells, TNF also acts on neutrophils to enhance their production of superoxide [Shalaby et al., 1985], induces the release of lysozymes and hydrogen peroxide and also induces neutrophil degranulation [Kiobbanoff et al., 1986].

TNF is known to be synthesized by a number of cell types, including cells of the haematopoietic and nonhaematopoietic lineages [Beutler, 1990; Spriggs et al., 1988; Jevnikar et al., 1991], although it is now recognized that macrophages are one of the major sources of TNF [Mannel et al., 1981]. Although macrophages do express receptors for TNF, little is known about the function of TNF on the macrophage although it has been shown to be chemotactic to these cells [Ming et al., 1987]. TNF is known to induce the production of GM-CSF by a number of cell lines and may cause the activation of macrophages indirectly via this pathway [Munker et al., 1986]. TNF also exerts an effect on lymphocytes. TNF acts to induce the up-regulation of TNF receptor (TNFR) on primary cultured T cells [Scheurich et al., 1987] and has also been shown to induce the expression of the HLA-DR antigen and the interleukin (IL)-2 receptor thus resulting in an enhanced proliferative response of T cells, particularly to IL-2. Conversely, TNF has been shown to inhibit pokeweed mitogen-induced proliferation and differation of B cells [Kashiwa et al., 1987].

Although this is by no means a comprehensive description of the intricate interactions that TNF is involved in, it can be clearly seen that as well as having direct cytotoxic effects, TNF also acts as an important immune modulator and is one of the primary initiators and modulators of the immune response.

TNF exists in solution as a homotrimer of 17-kD subunits each synthesized from a 26-kD transmembrane propeptide that itself appears to exhibit biological activity when membrane bound [Kreigler et al.,1988], whilst LT-α is a 25-kD soluble protein, synthesized primarily by lymphocytic cells [Paul and Ruddle, 1988]. Although antigenically distinct, it displays approximately 30% sequence homology with TNF and also close homology with the recently cloned 33kD glycosylated transmembrane protein LT-β [Browning et al., 1993]. The function of both LTs still remains unclear. LT-α is present on the surface of activated T cells, B cells and lymphokine-activated killer (LAK) cells heterodimerized to the LT-β protein. No function has yet been ascribed to these molecules although human peripheral blood mononuclear cells (PBLs) stimulated with either anti-CD3 antibody or IL-2 express mRNA for both LTs [Browning et al., 1993]. It is interesting that this heterodimeric molecule is incapable of binding to either of the two known TNF receptors (p55 and p75). Crowe et al. [1994] have recently identified a receptor specific for LT-β, however little is know about its function. Browning et al. [1993] hypothesize that the surface-bound LT-$\alpha\beta$ dimer may be important in cell-cell contact-dependent programmed cell death, but this still remains unproven, whilst Crowe et al. [1994] suggest that, since LT-α knock-out mice display a dysfunction in lymphoid organ development, yet $TNFR_{55}$ or $TNFR_{75}$ knock-out mice develop a normal lyphoid organ system, the LT-α/LT-β complex and the LT-βR may play an important role in immune development.

Two separate receptors have been identified for TNF, which also show a high degree of specificity for LT-α but a low degree of specificity for LT-β. These receptors, termed p55 and p75 (or sometimes p60 and p80), differ in both their molecular weight and also in the mechanism by which they mediate the biological effects of TNF [Brockhaus et al., 1990]. The p55 receptor is expressed on a wide range of cell types and expression of p55 seems to be sufficient to result in the cytotoxic action of TNF on tumour cells in vitro [Balkwill, 1993], independent of the expression of the p75 receptor [Higuchi and Aggarwal, 1994]. The p75 receptor shows a more specific pattern of expression and is found primarily on lymphoid and myeloid cells [Balkwill, 1993]. Whether or not the p75 receptor plays a function in the cytotoxicity of TNF is debatable, with a number of studies suggesting that in some cases this receptor is capable of inducing cytotoxicity upon cross-linking with specific antisera [Medvedev et al., 1994]. However, studies on transgenic mice deficient in p55 suggest that p75 is largely restricted to an accessory role as compared to p55 [Rothe et al., 1994].

TNF and Asthma

Asthma is characterized by an acute and chronic inflammatory response triggered by a large variety of specific and nonspecific agents. A characteristic of this response is degranulation of mast cells within the airways driven by the cross-linking of allergen-specific IgE molecules. This results in release of preformed factors including histamine, platelet activation factor (PAF), eicosanoids and several members of the cytokine family including TNF. There is now a large body of data which suggests that TNF may play a pivotal role in the resultant inflammatory response.

Elevated levels of TNF protein have been demonstrated in biopsies from the airways of asthmatic individuals. In addition, airway cells obtained at bronchoalveolar lavage (BAL) in asthma also demonstrate an increased ability to secrete TNF [Cembrzynska-Nowak et al., 1993; Broide et al., 1992; Gosset et al., 1991; Virchow et al., 1995]. Studies of peripheral monocytes following specific bronchoprovocation challenge in occupational asthma [Siracusa et al., 1992] and also in alveolar macrophages obtained at BAL [Maestrelli et al., 1995] show increased levels of TNF. LPS-stimulated monocytes from asthmatics produce 3-fold greater TNF than controls [Hallsworth et al., 1994]. Furthermore, increased TNF levels have been found in blood and sputum during asthmatic attacks.

There are several cellular sources and actions of TNF. TNF is increased in nasal polyps, primarily as a result of their high eosinophil content [Finotto et al., 1994], and is increased 7-fold in mast cells obtained from asthmatic biopsies [Bradding et al., 1994]. TNF induces IL-6, IL-8 and GM-CSF in bronchial epithelial cells and in vitro studies have demonstrated that T cells from asthmatic individuals will adhere more readily to airway smooth muscle cells if the muscle cells have been pretreated with TNF. This effect is modulated via up-regulation of the adhesion molecules ICAM-1 and its ligand LFA-1 [Lazaar et al., 1994]. In addition, TNF from alveolar macrophages up-regulates ICAM-1 and ELAM-1 on endothelial cells [Lassalle et al., 1991, 1993] and ICAM-1 on bronchial epithelial cells [Bloemen et al., 1993].

Polymorphisms within the TNF Locus

The TNF genes are located within the highly polymorphic region of human chromosome 6p termed the class III region of the MHC. Although the number of polymorphic regions identified to date within the human genome has not been formally calculated, there are at least 10 within the human TNF locus. These polymorphic regions are divided into two distinct types

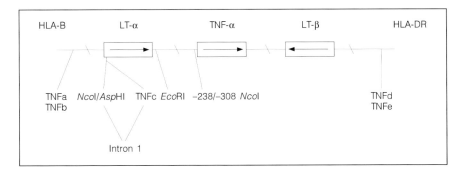

Fig. 1. The TNF locus: MHC class III region on chromosome 6p21.1–21.2 (see text for details).

comprising restriction fragment length polymorphisms (RFLP) (LT-α EcoRI RFLP, LT-α NcoI RFLP, –308 TNF-α RFLP, –238 TNFα RFLP and AspHI LT-α RFLP) [Partanen and Koskimies, 1988; Messer et al., 1991; Wilson et al., 1992; D'Alfonso and Richiardi, 1994; Ferencik et al., 1992] and five microsatellite polymorphisms (TNFa, TNFb, TNFc, TNFd and TNFe) [Jongeneel et al., 1991; Udalova et al., 1993]. These regions exhibit various degrees of polymorphisms and span the entire length of the TNF gene cluster (fig. 1).

The EcoRI RFLP described by Partanen and Koskimies [1988] is a biallelic polymorphism with the polymorphic restriction enzyme site located in the 3′ untranslated region of exon 4 of the LT-α gene. This RFLP produces two alleles defined by Southern hybridization of a common 2.4 kb and a rare 2.5 kb which was present in 4 of the 67 (6%) unrelated individuals in the original study. This RFLP shows no association with the common auto-immune-associated HLA antigens B8, DR3 or B27, but does show a strong association with B40 [Partanen and Koskimies, 1988]. Since the 2.5 kb allele of the EcoRI RFLP is so rare, little work has been performed regarding the function or associations of this polymorphism in disease.

The LT-α NcoI RFLP was initially defined by Southern hybridization with a TNF probe to give two alleles of 5.5 kb (termed the B*1 allele) and 10.5 kb (termed the B*2 allele) and for a number of years it was thought that the polymorphic NcoI site was within the TNF gene [Fugger et al., 1989]. Messer et al. [1991], however, mapped a 56.8 kb region of a number of overlapping genomic DNA clones and by direct sequencing located the polymorphic restriction site to within the first intron of the LT-α gene, at position 252. This polymorphism results in a guanine to adenine substitution to create the NcoI recognition site. Messer et al. [1991] also showed an associated amino acid substitution at position 26 of the LT-α gene with an asparagine associated with B*1 allele and threonine with the B*2 allele.

Functional analyses of the various LT-α RFLP genotypes have demonstrated an association with specific genotypes and the production of TNF in vitro. The most recent study by Pociot et al. [1993a] shows a distinct pattern of TNF secretion, but interestingly not LT-α secretion, associated with various LT-α RFLP genotypes. The authors report a significantly higher production of TNF by LPS-stimulated monocytes from individuals homozygous for the B*2 allele than those levels seen in B*1 homozygous individuals, whilst individuals heterozygous at this locus showed an intermediate level of TNF under the same stimulatory conditions. Although the LT-α RFLP is located within the gene for LT-α and is also associated with a variant amino acid within the LT-α protein, Pociot et al. [1993a] report no difference in the level of LT-α secretion by phytohaemagglutinin (PHA)-stimulated peripheral blood mononuclear cells (PBMCs). These results are in contrast to those reported by Messer et al. [1991], who reported an increased level of LT-α secretion associated with the TNFB*1 allele, in PHA-stimulated T cells. In the study by Messer et al. [1991] the authors report no association between TNF secretion in PHA-stimulated PBMCs from individuals homozygous for either of the two LT-α alleles, whilst individuals homozygous for the B*1 allele showed a significant increase in LT-α production over those individuals homozygous for the B*2 allele.

The –308 TNF-α RFLP is one of two RFLPs that have been identified, to date, within the gene for TNF. Like all RFLPs the –308 RFLP is a biallelic polymorphism describing two alleles, a common allele termed TNF1 and a rare allele termed TNF2 [Wilson et al, 1992]. This polymorphism results in the substitution of a guanine by an adenosine in the promotor region at position –308, relative to the site of initiation, in the TNF2 allele.

Both LT-α NcoI and –308 TNF RFLPs shows a strong association with alleles of the MHC. Wilson et al. [1993] have shown a strong association with the uncommon, TNF2 allele, and the extended HLA haplotype HLA-A1, -B8, -DR3. This extended haplotype has now been expanded to include the alleles of the LT-α RFLP as well as alleles of the TNFa, TNFb and TNFc microsatellites (see below). This haplotype now extends to DQB1*0201, DQA1*0501, DRB1*0301, TNF2, LT-α B*1, B8, A1/2, C4A *null* and has been observed in control individuals, in patients with insulin-dependent diabetes mellitus [Pociot et al., 1993b] and in patients with systemic lupus erythematosis [Wilson et al., 1994].

Constructs involving the two TNF allelic promotors upstream of a CAT reporter system result in a 6- to 7- fold increase in transcription in both PHA-stimulated and PHA-unstimulated Raji B-cell line transfected with the TNF-2 bearing promotor [Wilson, et al, 1994]. However, studies by a group in The Netherlands on similar CAT constructs have failed to show a similar effect [pers. commun.].

A recent report by D'Alfonso and Richiardi [1994] identifies a further RFLP within the promotor region of the TNF gene. It has been shown that changes in nucleotide sequence can result in changes in the superhelical structure of genomic DNA and that these changes can influence the electophoretic mobility [Calladine et al., 1988]. Using this knowledge D'Alfonso and Richiardi [1994] were able to detect a polymorphism distinct from the –308 TNF–α RFLP at position –238 of the TNF gene. This region of the promotor forms part of the highly conserved Y box which has been shown to be essential for MHC class II promotor function [Benoist and Mathis, 1990]. This polymorphism also forms part of an extended MHC haplotype and the authors report a number of associations including the rare TNFA-A allele with HLA-DR3.

A further RFLP has been reported within the same intron of the LT-α gene as the LT-α NcoI RFLP [Ferencik et al., 1992]. Like all of the RFLPs, this polymorphism results due to the substitution of a single base resulting in the formation or destruction of an AspHI restriction endonuclease recognition sequence. Like the EcoRI RFLP, there is very little further data available regarding this polymorphism. It has been reported that there is no linkage between the two RFLP loci in this intron or with alleles of the MHC [Ferencik et al., 1992].

Although the RFLPs within the TNF locus can provide invaluable evidence to the function of TNF in various diseases, their usefulness is limited due to their small degree of polymorphism. Five regions of dinucleotide repeat microsatellite polymorphism have been found distributed throughout the TNF locus [Nedospasov et al., 1991; Jongeneel et al., 1991; Udalova et al., 1993]. These loci show varying degrees of polymorphism.

The TNFa and TNFb polymorphisms consist of AC/GT and TC/GA repeat units respectively. These two loci are found within a 200 bp region located 3.5 kb upstream of the LT-α promotor and show 13 and 7 alleles respectively [Nedospasov et al., 1991; Jongeneel et al, 1991]. The TNFc polymorphism is biallelac and consists of TC/GA repeat units and is found within the first intron of the LT-α gene [Nedospasov et al., 1991]. The two most recently described microsatellite polymorphisms have been termed TNFd and TNFe and are located 8–10 kb downstream of the TNF gene. These polymorphisms both consist of TC/GA repeat units and have 7 and 3 alleles respectively [Udalova et al., 1993].

Pociot et al. [1993a] have shown an association between the TNFa2 allele and higher secretion of TNF by LPS-stimulated monocytes ($\sim$3.4 ng/ml) whilst TNFa6 was associated with a significantly lower level of secretion of TNF ($\sim$2.4 ng/ml), compared to a baseline taken as 3 ng/ml. TNFc2-positive monocytes also displayed a significantly higher level of TNF secretion, although the exact value for this was not presented. Like the LT-α RFLP, this study failed to show any association with TNFa or TNFc alleles and the secretion of LT-α by PHA-stimulated PBMCs [Pociot et al., 1993a].

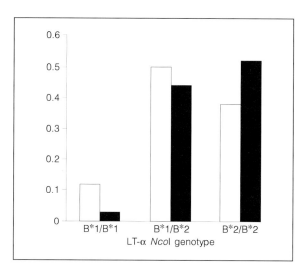

Fig. 2. Association of LT-α B*2/B*2 with wheeze (p = 0.03) in 224 subjects. □ = Non-asthmatic; ■ = asthmatic (see text for details).

Like the RFLPs within the TNF locus, the alleles of the five microsatellites seem to form part of an extended MHC haplotype. Four such haplotypes have been identified which correlate with TNF secretory capacity of LPS-stimulated monocytes [Pociot et al., 1993a]. Analysis of 101 cell lines from the Human Leukocyte Antigen Workshop reference panel for alleles of the five microsatellite polymorphisms and the LT-α, *Nco*I RFLP have identified a number of common extended MHC haplotypes (for a complete description of these haplotypes, see Udalova et al. [1993]). The authors report a number of unusual differences between the cell lines. Whilst a number of cell lines were shown to be identical for their extended haplotype, a number of lines extended haplotypes were created by recombination, which in one case occurred between the TNFb microsatellite locus and the HLA-B locus. Other cases were identified where cell lines shared the same HLA serological haplotype but showed differing TNF alleles. These cell lines have been used as control for HLA typing due to the fact that the majority of them are homozygous for the serologically defined HLA alleles. On analysis of the TNF alleles of several of these cell lines, Udalova et al. [1993] report that some of these lines show heterozygosity for alleles of the TNF locus. Those cell lines which showed heterozygosity for one of the TNF alleles, also showed heterozygosity for one of the HLA alleles. It is now clear that alleles within the TNF locus show a high degree of association with alleles of the HLA system. The ability to subdivide extended HLA haplotypes by incorporating haplotypes of the TNF

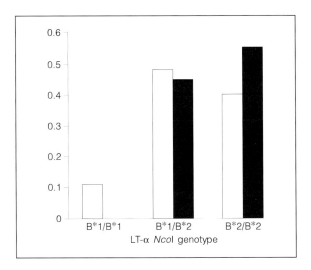

Fig. 3. Random selection from 2,514 subjects: 224 selected, of which 13% asthma and 35% atopic. LT-α B*1/B*1 associated with protection from asthma (p = 0.08). □ = Nonasthmatic; ■ = asthmatic (see text for details).

polymorphisms may allow a more detailed analysis of the association of the MHC and in particular TNF with human diseases.

TNF Polymorphisms and Asthma

We examined whether the B*2/B*2 homozygote genotype at the LT-α *Nco*I locus has a higher, or the B*1/B*1 has a lower incidence of asthma than other genotypes. We selected a random cohort of 224 subjects derived from a panel of 2,415 individuals randomly ascertained from the electoral register. This cohort comprised 100 atopic, 33 asthmatic and 91 nonatopic nonasthmatic subjects which approximated to the frequency of this disease in the larger cohort. Atopy was defined as a raised serum total IgE or a positive skin prick test reaction to either cat, grass mix or house dust mite $\geq - 1$ mm when compared to a negative control. Asthma was considered to be present only when diagnosed by a physician and bronchial hyperreactivity defined as a PD_{20} of < 12.25 μmol methacholine.

Oligonucleotide primers were synthesized according to the published sequence of Messer et al. [1991]. A 750 bp fragment containing the polymorphism was amplified by the polymerase chain reaction and was subjected to restriction endonuclease digestion (fig. 1). Fragments were analysed on a 1% agarose gel.

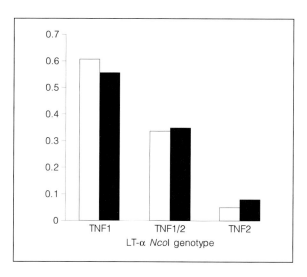

Fig. 4. Bronchial reactivity (–308 TNF) in 558 subjects in association with bronchial hyperreactivity (p = 0.038) and no association with asthma, atopy or IgE. □ = not reactive; ■ = reactive (see text for details).

We were able to demonstrate a significant protection of B*1/B*1 genotype individuals against recurrent wheeze (p = 0.03; fig. 2) and a borderline protection against asthma (p = 0.08; fig. 3). However, there was no association with bronchial reactivity or atopy. In view of this preliminary result we studied two further cohorts of subjects. Firstly, a larger cohort of 566 subjects from our random population and secondly 280 subjects from 60 nuclear families containing at least 1 asthmatic subject. The 506 random subjects comprising 157 normal, 161 atopic nonreactive, 158 reactive and atopic and 90 nonatopic nonreactive individuals. However, in both cohorts we showed no association between any LT-α genotype and asthma or atopy phenotype. From these studies we concluded that LT-α NcoI alleles were likely to be a component of the genetic predisposition to asthma in only a small proportion of subjects.

Following preliminary data from Moffat and co-workers [pers. commun.] of an association between genotypes at the –308 TNF polymorphism and asthma, we analysed this marker in our random cohort. The –308 TNF RFLP genotype was analysed using the protocol of Wilson et al. [1992]. Unlike the LT-α RFLP, we showed no association between either wheeze or asthma. However, we found an association between individuals homozygous for allele two of this polymorphism and bronchial hyperreactivity (p = 0.038; fig 4). We further examined this association using logistic regression analysis to determine whether other phenotypes may influence the association. We found that the association was strengthened by correcting for baseline FEV_1.

Conclusions and Future Work

Given the evidence for a role of TNF in the pathogenesis of asthma and our initial findings, we feel these observations need expanding and clarifying to encompass the other TNF polymorphisms. Given the intricacy of the associations between polymorphisms of not only this locus but also of the known associations between this locus and other regions of the MHC, there is still much work to be done to fully determine whether this locus has a major role to play in contributing to the genetic component of asthma. It is clear that larger studies may provide further data to support or refute our initial findings.

References

Aiyer R, Aggarwal BB: Tumour necrosis factors; in Podock E (ed): CRC Handbook of Cytolytic Lymphocytes and Complement as Effectors of the Immune System. Boca Raton, CRC Press, 1987.

Balkwill FR: Improving on the formula. Nature 1993;361:206–207.

Benoist C, Mathis D: Regulation of major histocompatibility complex class II genes: X, Y and other letters of the alphabet. Annu Rev Immunol 1990;8:681–715.

Bertolini DR, Nedwin G, Bringman T: Stimulation of bone resorption and inhibition and bone formation in vitro by human tumour necrosis factor. Nature 1986;319:516–518.

Beutler B: Cachectin/tumor necrosis factor and lymphotoxin; in Sporn MB, Roberts AB (eds): Peptide Growth Factors. II. Berlin, Springer 1990, pp 39–70.

Beutler B, Milsark IW, Cerami AC: Passive immunization against cachectin/tumour necrosis factor protects mice from lethal effect of endotoxin. Science 1985;229:869–871.

Bloeman PG, van den Tweel MC, Henricks PA, et al: Expression and modulation of adhesion molecules on human bronchial epithelial cells. Am J Respir Cell Mol Biol 1993;9:586–593.

Bradding P, Roberts JA, Britten KM, et al: IL-4, -5 and -6 and TNF-α in normal and asthmatic airways: Evidence for the human mast cell as a source of these cytokines. Am J Respir Cell Mol Biol 1994;10:471–480.

Brockhaus M, Schoenfeld H-J, Schlaeger E-J, et al: Identification of two types of tumour necrosis factor receptors on human cell lines by monoclonal antibodies. Prod Nat Acad Sci USA 1990;87:3127–3131.

Broide DH, Lotz M, Cuomo AJ, et al: Cytokines in symptomatic asthma airways. J Allergy Clin Immunol 1992;89:958–967.

Browning JL, Ngam-ek A, Lawton P, et al: Lymphotoxin β, a novel member of the TNF family that forms a heteromeric complex with lymphotoxin on the cell surface. Cell 1993;72:847–856.

Calladine CR, Drew HR, McCAll MJ: The intrinsic curvature of DNA in solution. J Mol Biol 1988;201:127–137.

Carswell EA, Old LJ, Kassel RC, et al: An endotoxin-induced serum factor that causes necrosis of tumors. Proc Nat Acad Sci USA 1975;72:3666–3670.

Cembrzynska-Nowak M, Szklarz E, Inglot AD, Teodorczyk-Injeyan JA: Elevated release of TNF-α and IFN-γ by bronchoalveolar leukocytes from patients with bronchial asthma. Annu Rev Respir Dis 1993;147:291–295.

Coley WB: Contribution to the knowledge of sarcoma. Ann Surg 1891;14:199–220.

Crowe PD, van Arsdale TL, Walter BN, et al: A lymphotoxin-beta-specific receptor. Science 1994;264: 707–710.

D'Alfonso S, Richiardi PM: A polymorphic variation in a putative regulation box of the TNFA promoter region. Immunogenetics 1994;39:150–154.

Dayler JM, Beutler B, Cerami A: Cachectin/tumor necrosis factor stimulates collagenase and PGE_2 production by human synovial cells and dermal fibroblasts. J Exp Med 1985;162:2163–2168.

Ferencik S, Lindermann M, Horsthemke B, et al: A new restriction fragment length polymorphism of the human TNF-β gene detected by AspHI digest. Eur J Immunogenet 1992;19:425–430.

Finotto S, Marshall JS, Gauldie J, et al: TNF-α production by eosinophils in the upper airways (nasal polyposis). J Immunol 1994;153:2278–2289.

Fugger L, Morling N, Ryder LP, et al: NcoI restriction fragment length polymorphism of the tumour necrosis factor region in primary biliary cirrhosis and in health Danes. Scand Immunol 1989;30: 185–189.

Gamble JR, Harlan JM, Klebanoff SJ, et al: Stimulation of the adherence of neutrophils to umbilical vein endothelium by human recombinant tumor necrosis factor. Proc Natl Acad Sci USA 1985;82: 8667–8673.

Gosset P, Tsicopoulos A, Wallaert B, et al: Increased secretion of TNF-α and IL-6 by alveolar macrophages consecutive to the development of the late asthma reaction. J Allergy Clin Immunol 1991;88: 561–571.

Gray PW, Aggarwal BB, Benton CV, et al: Cloning and expression of CDNA for human lymphotoxin, a lymphokine with tumor necrosis activity. Nature 1984;312:721–724.

Hallsworth MP, Soh CP, Lane SJ, et al: Selective enhancement of GM-CSF, TNF-α, IL-1β and IL-8 production by monocytes and macrophages in bronchial asthma. Eur Respir J 1994;7:1096–1102.

Higuchi M, Aggarwal BB: Overlapping and nonoverlapping functions of p60 and p80 forms of the human tumor necrosis factor receptors. Eur Cytokine Netw 1994;5:108.

Jevnikar AM, Brennan DC, Singer GG, et al: Stimulated kidney tubular epithelial cells express membrane associated and secreted TNF-α. Kidney Int 1991;40:203–211.

Jongeneel CV, Briant L, Udalova IA, et al: Extensive genetic polymorphism in the human tumor necrosis factor region and relation to extended HLA haplotypes. Proc Natl Acad Sci USA 1991;88:9717–9721.

Kashiwa H, Wright SC, Bonavida B: Regulation of B-cell maturation and differentiation. I. Suppression of pokeweed mitogen-induced B cell differentiation by tumour necrosis factor. J Immunol 1987; 138:1383–1390.

Kiobbanoff SJ, Vadas MA, Harlan JM, et al: Stimulation of neutrophils by tumor necrosis factor. J Immunol 1986;136:4220–4225.

Kriegler M, Pererez C, DeFay K, et al: A novel form of TNF/cachectin is a cell surface cytotoxic transmembrane protein: Ramifications for the complex physiology of TNF. Cell 1988;53:45–53.

Lassalle P, Delneste Y, Gosset P, et al: Potential implication of endothelial cells in bronchial asthma. Int Arch Allergy Appl Immunol 1991;94:233–238.

Lassalle P, Gosset P, Delneste Y, et al: Modulation of adhesion molecule expression on endothelial cells during the late asthmatic reaction: Role of macrophage-derived TNF-α. Clin Exp Immunol 1993; 94:105–110.

Lazaar AL, Albelda SM, Pilewski JM, et al: T lymphocytes adhere to airway smooth muscle cells via integrins and CD44 and induce smooth muscle DNA synthesis. J Exp Med 1994;180:807–816.

Maestrelli P, di Stefano A, Occari P, et al: Cytokines in the airway mucosa of subjects with asthma induced by toluene diisocyanate. Am J Respir Crit Care Med 1995;151:607–612.

Mannel DN, Falk W, Meltzer MS: Inhibition of nonspecific tumoricidal activity by activated macrophages with antiserum against a soluble cytotoxic factor. Infect Immunol 1981;33:156–164.

Medvedev AE, Sundan A, Espevik T: Involvement of the tumor necrosis factor receptor p75 in mediating cytotoxicity and gene regulating activities. Eur J Immunol 1994;24:2842–2849.

Messer G, Spengler U, Jung MC, et al: Polymorphic structure of the tumor necrosis factor (TNF) locus: An NcoI polymorphism in the first intron of the human TNF-β gene correlates with a variant amino acid in position 26 and a reduced level of TNF-β production. J Exp Med 1991;173:209–219.

Ming WJ, Bersani L, Mantovani A: Tumor necrosis factor is chemotactic for monocytes and polymorpho-nuclear leukocytes. J Immunol 1987;138:1469–1474.

Munker R, Gasson J, Ogawa M, et al: Recombinant human TNF induces production of granulocyte-monocyte colony-stimulating factor. Nature 1986;323:79–82.

Nedospasov SA, Udalova IA, Kuprash DV, et al: DNA sequence polymorphism at the human tumor necrosis factor (TNF) locus: Numerous TNF/lymphotoxin alleles tagged by two closely linked microsatellites in the upstream region of the lymphotoxin (TNF-β) gene. J Immunol 1991;147: 1053–1059.

Partanen J, Koskimies S: Low degree of DNA polymorphism in the HLA-linked lymphotoxin (tumour necrosis factor β) gene. Scand J Immunol 1988;28:313–316.

Paul NL, Ruddle NH: Lymphotoxin. Annu Rev Immunol 1988;6:407–438.

Pennica D, Nedwin GE, Hayflick JS, et al: Human tumor necrosis factor: Precursor structure, expression, and homology to lymphotoxin. Nature 1984;312:724–729.

Pociot F, Briant L, Jongeneel CV, et al: Association of tumor necrosis factor (TNF) and class II major histocompatibility complex alleles with the secretion of TNF-α and TNF-β by human mononuclear cells: A possible link to insulin-dependent diabetes mellitus. Eur J Immunol 1993a;23:224–231.

Pociot F, Wilson AG, Nerup J, et al: No association between a tumor necrosis factor-α promoter region polymorphism and insulin-dependent diabetes mellitus. Eur J Immunol 1993b;23:3043–3049.

Rothe J, Mackag F, Bluetmann H, et al: Phenotypic analysis of TNFR1-deficient mice and characterization of TNFR1-deficient fibroblasts in vitro. Circ Shock 1994;44:51–56.

Scheurich P, Thoma B, Ucer U, et al: Immunoregulatory activity of recombinant tumor necrosis factor (TNF)-alpha: Induction of TNF receptors on human T cells and TNF-alpha-mediated enhancement of T cell responses. J Immunol 1987;138:1786–1790.

Shalaby MR, Aggarwal BB, Rinderknecht E, et al: Activation of human polymorphonuclear neutrophil function by gamma interferon and tumor necrosis factor. J Immunol 1985;135:2069–2073.

Siracusa A, Vecchiarelli A, Brugnami G, et al: Changes in IL-1 and TNF production by peripheral blood monocytes after specific bronchoprovocation tests in occupational asthma. Am Rev Respir Dis 1992;146:408–412.

Spriggs DR, Imamura K, Rodriguez C, et al: Tumour necrosis factor expression in human epithelial tumor cell lines. J Clin Invest 1988;81:455–460.

Sugarman BJ, Aggarwal BB, Hass PE, et al: Recombinant human tumor necrosis factor-α: Effects on proliferation of normal and transformed cells in vitro. Science 1985;230:943–946.

Tracey KJ, Beutler B, Lowery SF: Shock and tissue injury induced by recombinant human cachectin. Science 1986;243:470–474.

Udalova IA, Nedospasov SA, Webb GC, et al: Highly informative typing of the human TNF locus using six adjacent polymorphic markers. Genomics 1993;16:180–186.

Virchow JC Jr, Walker C, Hafner D, et al: T cells and cytokines in bronchoalveolar lavage fluid after segmental allergen provocation in atopic asthama. Am J Respir Crit Care Med 1995;151:960–968.

Wilson AG, de Vries N, Pociot F, et al: An allelic polymorphism within the human tumor necrosis factor-α promoter region is strongly associated with HLA A1, B8 and DR3 alleles. J Exp Med 1993;177: 557–560.

Wilson AG, di Giovine FS, Blakemore AIF, et al: Single base change in the human tumor necrosis factor (TNFα) gene detectable by NcoI restriction of PCR product. Hum Mol Genet 1992;1:353.

Wilson AG, Symons JA, McDowell TL, et al: Effects of a tumour necrosis factor (TNFα) promoter base 167 transition on transcriptional activity. Br J Rheumatol 1994;33:105.

D.A. Campbell, PhD, Molecular Medicine Unit, Clinical Sciences Building,
St. James' University Hospital, Leeds LS2 9JT (UK)

Hall, IP (ed): Genetics of Asthma and Atopy.
Monogr Allergy. Basel, Karger, 1996, vol 33, pp 138–152

....

Regulation of Cytokine Genes Implicated in Asthma and Atopy

David J. Cousins, Dontcho Z. Staynov, Tak H. Lee

Department of Allergy and Respiratory Medicine, United Medical and
Dental Schools, Guy's Hospital, London, UK

Introduction

The role of cytokines in the inflammatory processes observed in asthma
and atopy has become an area of intense research in recent years. This research
has demonstrated that a large number of cytokine genes are overexpressed by
inflammatory cells, especially T lymphocytes, in patients with atopic disease.
T-helper lymphoctyes have been divided into subsets based upon the cytokine
genes that they express upon activation [1]. T-helper (Th) 1 cells express
interleukin (IL)-2 and interferon-γ (IFNγ) but not IL-4 or IL-5, whereas Th2
cells express IL-4, IL-5, IL-10, and IL-13 but not IL-2 or IFNγ. Both cell
types express IL-3 and granulocyte-macrophage colony-stimulating factor
(GM-CSF). A third subset of cells termed Th0 has also been identified which
express all of the aforementioned cytokines. These T-cell subsets were originally
identified in the mouse and a similar, though not identical, pattern of cytokine
expression has since been observed in human cells [2]. This chapter will concern
the evidence for the overexpression of a Th2-like cytokine profile in atopy,
namely IL-4, IL-5 and GM-CSF, and the mechanisms by which these genes
are regulated. However it has now become clear that these cytokines are not
exclusively produced by T cells and that several other cell types express them,
this will also be discussed.

Involvement of IL-4, IL-5 and GM-CSF in Atopic Disease

Several research groups have studied the expression of these cytokines in
atopy and asthma, initially by T-cell cloning and more recently by immuno-

histochemistry and in situ hybridization. Evidence for a Th2-like cytokine profile in atopic disease first arose from T-cell cloning experiments in which *Dermatophagoides pteronyssinus* (DP) was used as a specific antigen to generate T cells from the peripheral blood of patients with atopic dermatitis [3]. Clones derived in this way produced IL-4 but no IFNγ as opposed to clones reactive to *Candida albicans* which had a Th1-like phenotype. Similarly a Th2 phenotype was observed using DP and *Lolium perenne* group I as antigens to derive T-cell clones from atopic patients [4]. Subsequent studies in which T-cell clones have been derived from disease sites have also shown a mainly Th2-like cytokine profile. Infiltrating T cells in patients with vernal conjunctivitis produce IL-4 but little IFNγ [5] and mostly Th2-like clones were derived from the skin of patients with atopic dermatitis [6, 7]. Also most allergen-specific CD4+ T-cell clones derived from bronchial biopsy specimens from patients with grass pollen-induced asthma displayed a Th2-like phenotype [8].

Studies in which bronchoalveolar lavage (BAL) samples from atopic asthmatics have been analyzed have demonstrated increased levels of IL-4, IL-5 and GM-CSF in the lavage fluid [9–11]. Interestingly the levels of IL-5 correlated with the extent of eosinophil infiltrate in the BAL suggesting that IL-5 may be responsible for the eosinophilia characteristic of atopic asthma [9]. Using in situ hybridization and immunochistochemistry, Robinson et al. [12] confirmed that a predominantly Th2-like T-cell population exists in the BAL from atopic asthmatics with an increase in T cells positive for IL-4, IL-5 and GM-CSF messenger RNA (mRNA). It is also interesting to note that treatment with corticosteroids, which alleviates the symptoms of asthma, causes a decrease in the numbers of BAL cells positive for IL-4 and IL-5 mRNA [13].

Other research using in situ hybridization and immunohistochemistry has also shown a predominantly Th2-like cytokine profile in biopsy specimens from atopic patients. mRNA for IL-3, IL-4, IL-5 and GM-CSF was detected by in situ hybridization of skin biopsies from allergen-induced late-phase cutaneous reactions in atopic subjects although the cellular sources were not analysed [14]. Several studies have demonstrated increased mRNA and protein production of IL-4, IL-5 and GM-CSF in bronchial biopsies from atopic asthmatics and rhinitics [15–20]. Interestingly the source of these cytokines in biopsy specimens was not exclusively T cells with mast cells [17, 21–23], eosinophils [20, 22, 24] and epithelial cells [17, 19, 25, 26] also expressing these genes. Recently, Kay et al. [18] have reported that the predominant cell type (70%) transcribing IL-4 and IL-5 mRNA in atopic asthmatics and rhinitics is CD3+ T cells with the remaining cells being eosinophils and mast cells.

Using a genetic approach, two groups have recently demonstrated linkage of several markers on chromosome 5 to a gene regulating total serum IgE levels in atopics [27, 28], particularly close linkage was observed to the IL-4

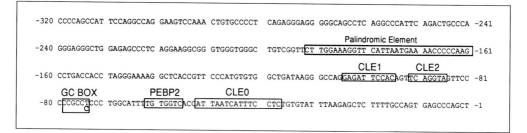

Fig. 1. Regulatory elements of the human GM-CSF promoter [84]. Sequence differences in elements found in the mouse promoter are shown below the human sequence. See text for definitions of labels.

gene in one of these studies [27]. This information suggests that a defect in the regulated expression of these genes may be a factor in atopic disease, therefore it is crucial to understand the mechanisms by which IL-4, IL-5 and GM-CSF expression is controlled.

Regulation of IL-4, IL-5 and GM-CSF Gene Expression in T Cells

GM-CSF, IL-4 and IL-5 are located in a small segment (5q23–31) on the long arm of human chromosome 5 and are in close proximity to IL-3, IL-13, interferon regulatory factor 1 (IRF-1) and several other as yet unidentified genes [29–42]. This suggests a common origin and/or some common regulatory mechanisms for expression of these genes. Whilst the function of these cytokine genes has been extensively investigated, our knowledge of how they are regulated, especially in human cells, is limited and only GM-CSF has been studied in detail.

GM-CSF Transcriptional Regulation

Over the last 5 years several groups have reported *cis*-acting elements in the promoter of GM-CSF that are active in T cells (fig. 1). Gasson and coworkers [43, 44] identified a positive element from −47 to −29 base pairs (bp) which is tissue-specific and is active in Jurkat cells upon activation with phytohaemaglutinin (PHA) and phorbol myristate acetate (PMA). The −47 to −29 region contains a 15-bp sequence ATTAATCATTTCCTC 100% conserved between mouse and human GM-CSF which differs by only one nucleotide (underlined, T in IL-5) from a corresponding region in the IL-5 promoter. Shannon et al. [45] have identified two protein-binding motifs. The first, Cytokine-1 (CK-1, binds NF-GMa), is a 10-bp sequence GAGATTCCAC, −105 to −96 bp from the transcription initiation site. The second element, cytokine-2 (CK-2, binds NF-GMb), is a 7-bp sequence TCAGGTA −92 to −86 bp from the transcription initiation site. Both of these elements were shown to bind specific nuclear proteins from the T-cell line HUT78 whether or not the cells were stimulated to express GM-CSF. These proteins are not present in cells which cannot express GM-CSF. This implies that these factors define tissue specificity and not an activation pathway. By performing DNase I hypersensitive site mapping, Cockerill et al. [46] have also identified an enhancer element 3 kilobases upstream of the GM-CSF gene. It contains four binding sites for the transcription factor AP-1, three of which also associate with NF-AT. They suggest that this enhancer is required for complete regulated activation of both GM-CSF and IL-3 in T cells. Using a transient transfection assay, Kaushansky [47] identified an additional positive regulatory element between −58 and −44 bp. This element was required for the up-regulation of reporter gene expression in response to ConA/PMA in T cells.

The most extensive work on the GM-CSF promoter has been carried out by Arai and co-workers [48–55]. They have used the mouse promoter with the human T-cell line Jurkat in numerous transfection, in vitro transcription, footprinting and gel retardation experiments. They have identified four *cis* elements responsible for efficient transcription. CLE1 (mouse −108 to −99, identical to human CK-1 −105 to −96), CLE2 (mouse −94 to −88, identical to human CK-2 −92 to −86), a GC box (mouse −84 to −73, corresponding to human −79 to −74) and CLE0 (mouse −54 to −40, corresponding to human −52 to −38). The CLE1 motif is required for induction of GM-CSF transcription by the p40^x transactivating protein (from HTLV-1) but has no effect on PMA/A23187 induction [48]. Both the CLE2 and GC box motifs are required together for efficient activation by PMA/A23187 or p40^x [49]. They have also shown that the region spanning CLE2/GC box contains two protein-binding motifs, the GM2 sequence GGTAGTTCCC, which binds the PMA/A23187-inducible factor NF-GM2, and CCGCCC which binds constitutive factors A$_1$, A$_2$ and B [50]. NF-GM2 appears to be NF-κB, or a very closely related factor [51, 56], and purified A$_1$ was identified as the transcription factor Sp1 [50, 55]. The human GM2 sequence is identical to the mouse, but the GC box has C to T mutation (CCGCCT), this may be important since a mutation from C to A abolishes binding of all three constitutive factors and also causes loss of transcriptional inducibility suggesting that Sp1 may not be involved in the regulation of human GM-CSF [50, 55]. It is interesting to note that they have reported that this element is also involved in the regulation of IL-3 [53]. In addition to the CLE2/GC box domain, they have also identified the CLE0 element [52]. This region is the same as that previously identified by Nimer et al. [44] and Heike et al. [54] who reported that it was required for PMA/A23187 induction of GM-CSF. Miyatake et al. [52] have shown that this region binds two factors NF-CLE0a and NF-CLE0b and that this complex is induced several-fold upon activation with PMA/A23187. The recognition sequences of these factors overlap; NF-CLE0b binds TCATTTCC (human -47 to -40), NF-CLE0b binds TTAATCAT (human -51 to -44). They have recently published data suggesting that NF-CLE0a resembles NF-AT and that NF-CLE0b is related to AP-1 [57, 58]. Both of these transcription factors are required for IL-2 activation [59]. Interestingly this CLE0 element is very highly conserved in several genes, notably IL-4, IL-5 and G-CSF. The proteins NF-CLE0a and b can bind to the CLE0 element from IL-5 and IL-4, but not G-CSF which is not produced by T cells [52]. This suggests that the CLE0 element plays a role in the coordinate induction of these cytokines in T cells.

A very recent paper by Takahashi et al. [60] has reported a further regulatory element, termed PEBP2 (-64 TGTGGTC -58 bp, human -62 to -56 bp), adjacent to the CLE0 element in the murine GM-CSF promoter. This element binds transcription factors of the polyoma virus enhancer-binding protein 2 (PEBP2) gene family. Different members of the PEBP2 family have different effects upon transcription of GM-CSF and they have postulated that the activity of this element is dependent upon the relative ratios of these proteins in the nucleus. This is however a recent discovery and considerable work is required to determine the importance of this element in the regulation of GM-CSF.

Recently we have reported a novel regulatory element in the promoter of the human GM-CSF gene [61]. It contains two symmetrically nested inverted repeats (-192 <u>CTTGG</u>AAAGG<u>TTCATT</u> <u>AATGAA</u>AACCC<u>CCAAG</u> -161). In transfection assays using the human GM-CSF promoter this element has a strong positive effect on the expression of a reporter gene by the human T-cell line Jurkat J6 upon stimulation with phorbol dibutyrate (PDBu) and ionomycin or anti-CD3. In DNA band-retardation assays this sequence produces six specific bands which involve one or other of the inverted repeats. We have also shown that a DNA-protein complex can be formed involving both repeats and probably more than one protein. The external inverted repeat contains a core sequence CTTGG...CCAAG which is also present in the promoters of several other T-cell expressed human cytokine genes including IL-4 and IL-5 (shown in figures 2 and 3, labelled 'palindromic element'). However the palindromic elements in these genes are larger than the core sequence suggesting that some of the interacting proteins may be different for different genes and may represent a family of novel transcription factors. Since this element is present in the promoters of all human Th2-like genes, it may

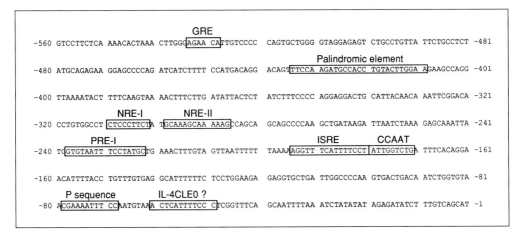

Fig. 2. Regulatory elements of the human IL-4 promoter [85]. See text for definitions of labels.

contribute to the coordinated expression of this group of cytokines. Whilst this element is present in the mouse and human promoters of the GM-CSF gene, which is expressed by all T cells, it is only present in the promoters of human and not mouse IL-4, IL-5 and IL-13 genes which are expressed by Th2-like cells. This suggests that there may be differences in the way in which these genes are regulated in human and mouse T cells.

IL-4 Gene Regulation

Although the IL-4 promoter region is relatively conserved from mouse to human, it differs considerably from the promoters of IL-5 and GM-CSF (fig. 2). It contains a putative glucocorticoid response element (AGAACA) and only a short fragment, TCATTT, of the CLE0 element present in IL-5 and GM-CSF is conserved. Transfection studies on the human IL-4 promoter have identified several *cis*-acting regions, the most proximal of which is the positive (P) sequence (-79 CGAAAATTCC -69) [62]. This region is required for activation of IL-4 in the T-cell line Jurkat and binds a factor NF(P) which appears to be very closely related to NF-κB, NF-AT [63] and also NF-Y [64], a factor required for expression of MHC class II genes. Two elements adjacent to each other have been reported between -195 and -172 bp (AGGTTTCATTTTCCTATTGGTCTG) [64]. These are an interferon stimulatory response element (ISRE) between -195 and -181 bp, and a CCAAT motif between -180 and -176 bp. The ISRE element is associated with IRF-2, a repressor, and an NF-1-like factor. Mutations in the ISRE element increase

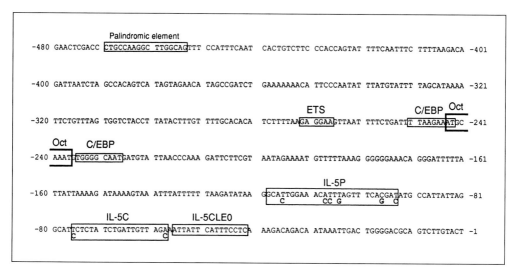

Fig. 3. Regulatory elements of the human IL-5 promoter [86]. Sequence differences in elements found in the mouse promoter are shown below the human sequence. See text for definitions of labels.

transcription suggesting that it acts as a negative element. The CCAAT element interacts with NF-Y and mutations in this element cause a decrease in transcription suggesting a positive role for NF-Y in IL-4 transcription. Another positive regulatory element (PRE-1, GTGTAATTTCCTATGC) is located between -239 and -224 bp [65]. This region binds two factors POS-1 and POS-2 which are both required to bind for full enhancer activity. These factors are ubiquitous and the enhancer is constitutively active showing no T-cell specificity.

Two negative regulatory elements (NRE) have also been identified; NRE-I (CTCCCTTCT) from -311 to -303 and NRE-II (GCAAAGCAAAAAG) from -300 to -288 [66]. In one study, it is reported that a T-cell-specific protein (Neg-1, binds to NRE-I) and a ubiquitous protein (Neg-2, binds to NRE-II) are associated with the two elements. In Jurkat cells, which are activated with PMA/ionomycin, the two NREs act in tandem to down-regulate the PRE-1 [65].

Unfortunately all of these studies have been performed using the Jurkat cell line which constitutively transcribes IL-4 and only shows limited up regulation of IL-4 upon activation. This is very different to the situation in a normal human T cell and therefore the activities of all of these elements will have to be validated in more suitable cells.

IL-5 Gene Regulation

Until now the data available on the transcriptional regulation of human IL-5 has been very limited due to the lack of a suitable cell line for use in transient transfection assays. However, several papers have recently been published which have used transient transfection assays to identify regulatory elements in the mouse IL-5 promoter. Using the mouse thymoma line EL-4, Lee et al. [67] have shown that dibutyryl cAMP (Bt$_2$cAMP) and PMA act synergistically to activate the mouse IL-5 promoter. They have identified four elements in the promoter that are required for complete activation of transcription [68]. The region designated IL-5A (-948 TTGAAAAGTGGGTCAA -933), a part of which has homology to NF-1-binding elements. Mutations in this region caused a 40% loss of promoter activity when activated with PMA and Bt$_2$cAMP. The sequence between -117 and -92 (GCACTG-GAAACCCTGAGTTTCAGGAC), known as IL-5P. Mutation of this region resulted in loss of 80% of the promoter activity and it appears to be an NF-AT-associated element. Two adjacent elements were also identified, IL-5C (-74 CCTCTATCTGATTGTTAGC -56) and IL-5CLE0 (-55 AAT-TATTCATTTCCTCAG -38) which is homologous to the GM-CSF CLE0 element. Both of these elements are essential for promoter activity since mutations in this region abolish transcription.

Naora et al. [69] have used a stable transfection technique with a mouse Th2 clone to show that the IL-5CLE0 region of the mouse promoter is required for functional activation. They have also shown that this region apparently binds proteins upon activation through different signaling pathways to those that activate GM-CSF. Bourke et al. [70] have confirmed that the IL-5A region is important for transcription of IL-5 and have shown that mutation of the NF-1 site actually causes constitutive transcription in their assay system [70]. This suggests that NF-1 acts as a negative regulator of IL-5 transcription.

We have been using DNase I footprinting to identify elements of the human IL-5 promoter that bind nuclear proteins derived from allergen-specific human T-cell clones [71]. This method has identified several regions which interact with transcription factors. The 15 nucleotide sequence (-56 AT-TATTCATTTCCTC -42 bp) which is homologous to CLE0 of GM-CSF. This element is fully conserved between mouse and human IL-5 and differs by only one nucleotide in GM-CSF (see above). The region -244 ATGCAAAT -237 bp which binds proteins of the Octamer (Oct) family is found in the promoters of many genes. These factors often work in concert with other nearby elements. In the IL-5 promoter this sequence is flanked by two CCAAT/enhancer-binding protein (C/EBP) elements (-235 TGGGGCAAT -227 bp) and (-251 TTAAGAAAT -243 bp). These sequences bind members of the C/EBP family such as nuclear factor-IL-6 (NF-IL-6) which has been shown to be involved in the activation of several cytokine genes [72, 73]. The motif GAGGAA between -272 and -267 bp, which binds transcription factors of the ets proto-oncogene family. It is different to the consensus sequence for the ETS1 factor found in activated T cells which is a negative regulator of IL-2 expression [74]. The function of these elements has yet to be determined in human T cells.

Regulation of IL-4, IL-5 and GM-CSF Gene Expression in Other Cell Types

Because of the relatively recent identification of mast cells, eosinophils and epithelial cells as sources of IL-4, IL-5 and GM-CSF in depth analysis of the transcriptional regulation of these genes has yet to be carried out. However, some data has recently been published regarding these cell types.

Mast Cells

Bradding et al. [21] have localized IL-4 to human mast cells derived from skin and respiratory tract and have shown that IL-4 exists in the cytoplasm of purified mast cells and that it is rapidly released upon cross-linking of the Fcε receptor with anti-IgE. This is a very different situation to that of T cells in which there is no pre-existing IL-4 mRNA or protein before activation occurs. This would suggest that constitutive transcription and translation of IL-4 occurs in mast cells and thus the regulatory mechanisms which control IL-4 production will be considerably different to those previously described for T cells. Indeed Plaut et al. [75] have shown that murine mast cells constitutively transcribe mRNA for IL-4 which is enhanced upon cross-linking of the Fcε receptor. Henkel et al. [76] have identified an enhancer element in the second intron of the murine IL-4 gene which is essential to the high levels of IL-4 expression observed in transformed murine mast cells. Prieschl et al. [77] have examined transient transfection of the human IL-5 promoter in the mouse mast cell line CPII and have found that an inducible NF-AT-like factor cooperates with a constitutive member of the GATA family in activation of IL-5 expression after the addition of IgE and antigen.

Eosinophils

It is now clear that eosinophils express IL-4, IL-5 and GM-CSF [20, 24, 78–81]. Because of the difficulty of obtaining large numbers of eosinophils, very little data has been published on the regulation of these genes in eosinophils. However, it is clear that peripheral blood eosinophils do not transcribe GM-CSF unless they are activated in vitro [79]. This is different to the regulation of IL-4 which has been shown to be constitutively transcribed and translated for storage in eosinophilic granules [81]. It is as yet unclear whether IL-5 is a constitutive product of eosinophils or if activation is required for transcription to occur.

Epithelial Cells

Several groups have shown that respiratory epithelial cells express GM-CSF both in vivo and in vitro [19,25,26,82]. It appears that expression of GM-CSF is constitutive but can be increased upon stimulation with various cytokines [26] and decreased by treatment with corticosteroids [19]. Marini et al. [25] have reported that epithelial cells do not transcribe mRNA for IL-5. No data are available regarding IL-4 expression by this cell type.

Conclusions

Overall the results detailed above show that the transcriptional regulation of IL-4, IL-5 and GM-CSF is a very complex process involving interactions between many *cis*-acting elements and the transcription factors that bind to them. It is important to note that the factor NF-AT appears to be required for the efficient activation of all of these genes in T cells by binding to various elements in their promoters. This is interesting since it is also essential for the activation of IL-2 [83], a Th1 cytokine. Similarly the palindromic element we have described is also present in the IL-2 promoter [61], suggesting that considerably more research is required in order to elucidate the mechanisms involved in the differential expression of Th1- and Th2-like cytokines by T-cell subsets and other cell types.

References

1 Mossmann TR, Coffan RL: T_{h1} and T_{h2} cells: Different patterns of lymphokine secretion lead to different functional properties. Annu Rev Immunol 1989;7:145–173.
2 Romagnani S: Human TH1 and TH2 subsets: Doubt no more. Immunol Today 1991;12:256–257.
3 Wierenga EA, Snock M, de Groot C, Chretien I, Bos JD, Jansen HM, Kapsenberg M: Evidence for compartmentalization of functional subsets of CD4+ T lymphocytes in atopic patients. J Immunol 1990;144:4651–4656.
4 Parronchi P, Macchia D, Piccinni M-P, Biswas P, Simonelli C, Maggi E, Ricci M, Asari AA, Romagnani S: Allergen- and bacterial antigen-specific T-cell clones established from atopic donors show a different profile of cytokine production. Proc Natl Acad Sci USA 1991;88:4538–4542.
5 Maggi E, Biswas P, Del Prete GF, Parronchi P, Macchia D, Simonelli C, Emmi L, De Carli M, Tiri A, Ricci M, Romanani S: Accumulation of Th2-like helper T cells in the conjunctiva of patients with vernal conjunctivitis. J Immunol 1991;146:1169–1174.

6   Van der Heijden FL, Wierenga EA, Bos JD, Kapsenberg ML: High frequency of IL-4-producing CD4+ allergen-specific T lymphoctyes in atopic dermatitis lesional skin. J Invest Dermatol 1991; 97:389–394.

7   Ramb-Lindhauer C, Feldmann A, Rotte M, Neumann C: Characterization of grass pollen reactive T-cell lines derived from lesional atopic skin. Arch Dermatol Res 1991;283:71–76.

8   Del Prete GF, De Carli M, D'Elios MM, Maestrelli P, Ricci M, Fabbri L, Romagnani S: Allergen exposure induces the activation of allergen-specific Th2 cells in the airway mucosa of patients with allergic respiratory disorders. Eur J Immunol 1993;23:1445–1449.

9   Walker, C, Bode E, Boer L, Hansel TT, Blaser K, Virchow J-C: Allergic and nonallergic asthmatics have distinct patterns of T cell activation and cytokine production in peripheral blood and broncho-alverolar lavage. Am Rev Respir Dis 1991;146:1'09–115.

10 Walder C, Bauer W, Braun RK, Menz G, Braun P, Schwarz F, Hansel TT, Villiger B: Activated T cells and cytokines in bronchoalveolar lavages from patients with various lung diseases associated with eosinophilia. Am J Respir Crit Care Med 1994;150:1038–1048.

11 Howell CJ, Pujol J-L, Crea AEG, Davidson R, Gearing AJH, Godard PH, Lee TH: Identification of an alveolar macrophage-derived activity in bronchial asthma that enhances leukotriene C4 generation by human eosinophils stimulated by ionophore A23187 as a granulocyte-macrophage colony-stimulating factor. Am Rev Respir Dis 1989;149:1340–1347.

12 Robinson DS, Hamid Q, Ying S, Tsicopoulos A, Barkans J, Bentley AM, Corrigan C, Durham SR, Kay AB: Predominant T_{h2}-like bronchoalveolar T-lymphocyte population in atopic asthma. N Eng J Med 1992;326:298–304.

13 Robinson D, Hamid Q, Ying S, Bentley A, Assoufi B, Durham S, Kay AB: Prednisolone treatment in asthma is associated with modulation of bronchoalveolar lavage cell interleukin-4, interleukin-5 and interferon-γ cytokine gene expression. Am Rev Respir Dis 1993;148:401–406.

14 Kay AB, Ying S, Varney VA, Gaga M, Dukrham SR, Moqbel R, Wardlaw AJ, Hamid Q: Messenger RNA expression of the cytokine gene cluster interleukin (IL)-, IL-4, IL-5 and granulocyte/macrophage colony-stimulating factor, in allergen-induced late-phase cutaneous reactions in atopic subjects. J Exp Med 1991;173:775–778.

15 Durham SR, Ying S, Varney VA, Jacobson MR, Sudderick RM, Mackay IS, Kay AB, Hamid QA: Cytokine messenger RNA expression for IL-3, IL-4, IL-5 and granulocyte/macrophage colony-stimulating factor in the nasal mucosa after local allergen provocation: Relationship to tissue eosinophilia. J Immunol 1992;148:2390–2394.

16 Hamid Q, Azzawi M, Ying S, Moqbel R, Wardlaw AF, Corrigan CJ, Bradley B, Durham SR, Collins JV, Jeffery PK, Quint DJ, Kay AB: Expression of mRNA for interleukin-5 in mucosal bronchial biopsies from asthma. J Clin Invest 1991;87:1541–1546.

17 Ackerman V, Marini M, Vittori E, Bellini A, Vassali G, Mattoli S: Detection of cytokines and their cell sources in bronchial biopsy specimens from asthmatic patients: Relationship to atopic status, symptoms, and level of airway hyperresonsiveness. Chest 1994;105:687–696.

18 Kay AB, Ying S, Durham SR: Phenotype of cells positive for interleukin-4 and interleukin-5 mRNA in allergic tissue reactions. Int Arch Allergy Immunol 1995;107:208–210.

19 Sousa AR, Poston RN, Lane SJ, Nakhosteen JA, Lee TH: Detection of GM-CSF in asthmatic bronchial epithelium and decrease by inhaled corticosteroids. Am Rev Respir Dis 1993;147:1557–1561.

20 Nanaka M, Nonaka R, Woolley K, Adelroth E, Miura K, O'Byrne P, Dolovich J, Jordana M: Localization of interleukin-4 in eosinophils in nasal polyps and asthmatic bronchial mucosa. J Allergy Clin Immunol 1995;95:220.

21 Bradding P, Feather IH, Howarth PH, Mueller R, Roberts JA, Britten K, Bews JPA, Hunt TC, Okayama Y, Heusser CH, Bullock GR, Church MK, Holgate ST: Interleukin-4 is localized to and released by human mast cells. J Exp Med 1992;176:1381–1386.

22 Bradding P, Feather IH, Wilson S, Bardin PG, Heusser CH, Holgate ST, Howarth PH: Immunolocalization of cytokines in the nasal mucosa of normal and perennial rhinitic subjects. J Immunol 1993; 151:3853–3865.

23 Bradding P, Roberts JA, Britten K, Montefort S, Djukanovic R, Mueller R, Heusser CH, Howarth PH, Holgate ST: Interleukin-4, -5 and -6 and tumor necrosis factor-α in normal and asthmatic

airways: Evidence for the human mast cell as a source of thee cytokines. Am J Respir Cell Mol Biol 1994;10:471–480.

24 Broide DH, Paine MM, Firestein GS: Eosinophils express interleukin-5 and granulocyte-macrophage colony-stimulating factor mRNA at sites of allergic inflammation in asthmatics. J Clin Invest 1992; 90:1414–1424.

25 Marini M, Vittori E, Hollemborg J, Mattoli S: Expression of the potent inflammatory cytokines, granulocyte-macrophage colony stimulating factor and interleukin-6 and interleukin-8, in bronchial epithelial cells of patients with asthma. J Allergy Clin Immunol 1992;89:1001–1009.

26 Cromwell O, Hamid Q, Corrigan CJ, Barkans J, Meng Q, Collins PD, Kay AB: Expression and generation of interleukin-8, IL-6 and granulocyte-macrophage colony-stimulating factor by bronchial epithelial cells and enhancement by IL-1β and tumour necrosis factor-α. Immunology 1992; 77:330–337.

27 Marsh DG, Neely JD, Breazeale DR, Ghosh B, Friedhoff LR, Ehrlich-Kautzky E, Schou C, Krishnaswamy G, Beaty TH: Linkage analysis of IL-4 and other chromosome 5q31.1 markers and total serum IrE concentrations. Science 1994;264:1152–1156.

28 Meyers DA, Postma DS, Panhuysen CIM, Xu J, Amelung PJ, Levitt RC, Bleecker ER: Evidence for a locus regulating total serum IgE levels mapping to chromosome 5. Genomics 1994;23:464–470.

29 Huebner K, Isobe M, Croce CM, Golde DW, Kaufman SE, Gasson JC: The human gene encoding GM-CSF is at 5q21-q32, the chromosome region deleted in the 5q- anomaly. Science 1985;230: 1282–1285.

30 Yang Y-C, Kovacic S, Kriz R, Wolf S, Clark SC, Wellems TE, Nienhuis A, Epstein N: The human genes for GM-CSF and IL-3 are closely linked in tandem on chromosome 5. Blood 1988;71:958– 961.

31 Sutherland GR, Baker E, Callen DF, Campbell HD, Young IG, Sanderson CJ, Garson OM, Lopez AF, Vadas MA: Interleukin-5 is at 5q31 and is deleted in the 5q-syndrome. Blood 1988;71:1150–1152.

32 Van Leeuwen BH, Martinson ME, Webb GC, Young IG: Molecular organization of the cytokine gene cluster, involving the human IL-3, IL-4, IL-5 and GM-CSF genes, on human chromosome 5. Blood 1989;73:1142–1148.

33 Huebner K, Nagarajan L, Besa E, Angert E, Lange BJ, Cannizzaro LA, van der Berghe H, Santoli D, Croce CM, Nowell PC: Order of genes on human chromosome 5q with respect to 5q interstitial deletions. Am J Hum Genet 1990;46:26–36.

34 Wasmuth JJ, Park C, Ferrell RE: Report of the committee on the genetic constitution of chromosome 5. Cytogenet Cell Genet 1989;51:137–148.

35 LeBeau MM, Lemons RS, Espinosa R III, Larson RA, Arai N, Rowley JD: Interleukin-4 and interleukin-5 map to human chromosome 5 in a region encoding growth factors and receptors and are deleted in myeloid leukemias with a del(5q). Blood 1989;73:647–650.

36 Chandrasekharappa SC, Rebelsky MS, Firak TA, LeBeau MM, Westbrook C: A long-range restriction map of the interleukin-4 and interleukin-5 linkage group on chromosome 5. Genomics 1990;6:94–99.

37 Warrington JA, Hall LV, Hinton LM Miller JN, Wasmuth JJ, Lovet M: Radiation hybrid map of 13 loci on the long arm of chromosome 5. Genomics 1991;11:701–708.

38 Warrington JA, Bailey SK, Armstrong E, Aprelikova O, Alitalo K, Dolganov GM, Wilcox AS, Sikela JM, Wolfe SF, Lovett M, Wasmuth JJ: A radiation hybrid of 18th growth factor, growth factor receptor, hormone receptor, or neurotransmittter receptor genes on the distal region of the long arm of chromosome 5. Genomics 1992;13:803–880.

39 Minty A, Chalon P, Derocq J-M, Dumont X, Guillemot J-C, Kaghad M, Labit C, Leplatois P, Liauzun P, Miloux B, Minty C, Casellas P, Loison G, Lupker J, Shire D, Ferrara P, Caput D: IL-13 is a new human lymphokine regulating inflammatory and immune responses. Nature 1993;362: 248–250.

40 Morgan JG, Dolganov GM, Robins SE, Hinton LM, Lovett M: The selective isolation of novel cDNAs encoded by the regions surrounding the human interleukin-4 and -5 genes. Nucleic Acids Res 1992;20:5173–5179.

41 Takahashi M, Yoshida MC, Satoh H, Hilgers J, Yaoita Y, Honjo T: Chromosomal mapping of the mouse IL-4 and human IL-5 genes. Genomics 1989;4:47–52.

42 Smirnov DV, Smirnova MG, Korobko VG, Frolova EI: Tandem arrangement of human genes for interleukin-4 and interleukin-13: Resemblance in their organization. Gene 1995;155:277–281.

43 Chan JY, Slamon DJ, Nimer SD, Golde DW, Gasson JC: Regulation of expression of human granulocyte-macrophage colony-stimulating factor. Proc Natl Acad Sci USA 1986;83:8669–8673.

44 Nimer SD, Morita EA, Martis MJ, Wachsman W, Gasson J: Characterization of the human granulocyte-macrophage colony-stimulating factor promoter region by genetic analysis: Correlation with DNase I footprinting. Mol Cell Biol 1988;8:1979–1984.

45 Shannon MF, Gamble JR, Vadas MA: Nuclear proteins interacting with the promoter region of the human granulocyte-macrophage colony-stimulating factor gene. Proc Natl Acad Sci USA 1988; 85:674–678.

46 Cockerill PN, Shannon MF, Bert AG, Ryan GR, Vadas MA: The granulocyte-macrophage colony-stimulating factor/interleukin-locus is regulated by an inducible cyclosporin A-sensitive enhancer. Proc Natl Acad Sci USA 1993;90:2466–2470.

47 Kaushansky K: Control of granulocyte-macrophage colony-stimulating factor production in normal endothelial cells by positive and negative regulatory elements. J Immunol 1989;143:2525–2529.

48 Miyatake S, Seiki M, DeWall Malefijt R, Heike T, Fujisawa J-1, Takebe Y, Nishida J, Shloami J, Yokota T, Yoshida M, Arai K-I, Arai N: Activation of T cell-derived lymphokine genes in T cells and fibroblasts: Effects of human T cell leukemia virus type 1 p40^x protein and bovine papilloma virus encoded E2 protein. Nucleic Acids Res 1988;16:6547–6566.

49 Miyatake S, Seiki M, Yoshida M, Arai K-I: T-cell activation signals and human T-cell leukemia virus type I-encoded p40^x protein activate the mouse granulocyte-macrophage colony-stimulating factor gene through a common DNA element. Mol Cell Biol 1988; 8:5581–5587.

50 Sugimoto K, Tsuboi A, Miyatake S, Arai K, Arai N: Inducible and non-inducible factors co-operatively activate the GM-CSF promoter by interacting with two adjacent DNA motifs. Int Immunol 1990;2:787–794.

51 Tsuboi A, Sugimoto K, Yodoi J, Miryatake S, Arai K, Arai N: A nuclear factor NF-GM2 that interacts with a regulatory region of the GM-CSF gene essential for its induction in response to T-cell leukemia line Jurkat cells and similarity to NF-κB. Int Immunol 1991;3:807–817.

52 Miyatake S, Shlomai J, Arai K-I, Arai N: Characterization of the mouse granulocyte-macrophage colony-stimulating factor (GM-CSF) gene promoter: Nuclear factors that interact with an element shared by three lymphokine genes – those for GM-CSF, Interleukin (IL)-4 and IL-5. Mol Cell Biol 1991;11:5894–5901.

53 Nishida J, Yoshida M, Arai K, Yokota T: Definition of a GC-rich motif as regulatory sequence of the human IL-3 gene: Coordinate regulation of the IL-3 gene by CLE2/GC box of the GM-CSF gene in T cell activation. Int Immunol 1991;3:245–254.

54 Heike T, Miyatake S, Yoshida M, Arai K, Arai N: Bovine papilloma virus encoded E2 protein activates lymphokine genes through DNA elements, distinct from the consensus motif, in the long control region of its own genome. EMBO J 1989;8:1411–1417.

55 Masuda ES, Yamaguchi-Iwai Y, Tsuboi A, Hung P, Arai K-I, Arai N: The transcription factor Sp1 is required for induction of the murine GM-CSF promoter in T cells. Biochem Biophys Res Commun 1994;205:1518–1525.

56 Schreck R, Baeuerle PA: NF-κB as inducible transcriptional activator of the granulocyte-macrophage colony-stimulating factor gene. Mol Cell Biol 1990;10:1281–1286.

57 Tokumitsu H, Masuda ES, Tsuboi A, Arai K-I, Arai N: Purification of the 120-kD component of the human nuclear factor of activated T cells (NF-AT). Biochem Biophys Res Commun 1993;196:737–744.

58 Masuda ES, Tokumitsu H, Tsuboi A, Shlomai J, Hung P, Arai K-I, Arai N: The granulocyte-macrophage colony-stimulating factor promoter cis-acting element CLE0 mediates induction signals in T cells and is recognized by factors related to AP1 and NFAT. Mol Cell Biol 1993;13:7399-7407.

59 Ullman KS, Northrop JP, Verweij CL, Crabtree GR: Transmission of signals from the T lymphocyte antigen receptor to the genes responsible for cell proliferation and immune function. Annu Rev Immunol 1990;8:421–452.

60   Takahashi A, Satake M, Yamaguchi-Iwai Y, Bae S-C, Lu J, Maruyama M, Zhang YW, Oka H, Arai N, Arai K-I, Ito Y: Positive and negative regulation of granulocyte-macrophage colony-stimulating factor promoter activity by *AML1*-related transcription factor, PEBP2. Blood 1995;86: 607–616.

61   Staynov DZ, Cousins DJ, Lee TH: A regulatory element in the promoter of the human granulocyte-macrophage colony-stimulating factor gene that has related sequences in other T-cell expressed cytokine genes. Proc Natl Acad Sci USA 1995;92:3606–3610.

62   Abe E, DeWaal Malefyt R, Matsuda I, Arai K, Arai N: An 11-base-pair DNA sequence motif apparently unique to the human interleukin-4 gene confers responsiveness to T-cell activation signals. Proc Natl Acad Sci USA 1992;89:2864–2868.

63   Matsuda I, Masuda ES, Tsuboi A, Behnam S, Arai N, Arai K-I: Characterization of NF(P), the nuclear factor that interacts with the regulatory P sequence (5'-AAAATTTCC-3') of the human interleukin-4 gene: Relationship to NF-κB and NF-AT. Biochem Biophys Res Commun 1994;199: 439–446.

64   Li-Weber M, Davydov IV, Frafft H, Drammer PH: The role of NF-Y and IRF-2 in the regulation of human IL-4 gene expression. J Immunol 1994;153:4122–4133.

65   Li-Weber M, Krafft H, Krammer PH: A novel enhancer element in the human IL-4 promoter is suppressed by a position-independent silencer. J Immunol 1993;151:1371–1382.

66   Li-Weber M, Eder A, Krafft-Czepa H, Krammer PH: T cell-specific negative regulation of transcription of the human cytokine IL-4. J Immunol 1992;148:1913–1918.

67   Lee HJ, Koyano-Nakagawa N, Naito Y, Nishida J, Arai N, Arai K-I, Yokota T: cAMP activates the IL-5 promoter synergistically with phorbol ester through the signaling pathway involving protein kinase A in mouse thymoma line EL-4. J Immulol 1993;151:6135–6142.

68   Lee HJ, Masuda ES, Arai N, Arai K-I, Yokota T: Definition of *cis*-regulatory elements of the mouse interleukin-gene promoter. J Biol Chem 1995;270:17541–17550.

69   Naora H, van Leeuwen BH, Bourke PF, Young IG: Functional role and signal-induced modulation of proteins recognizing the conserved TCATTT-containing promoter elements in the murine IL-5 and GM-CSF genes in T lymphocytes. J Immunol 1994;153:3466–3475.

70   Bourke PF, van Leeuwen BH, Campbell HD, Young IG: Localization of the inducible enhancer in the mouse interleukin-5 gene that is responsive to T-cell receptor stimulation Blood 1995;85: 2069–2077.

71   Cousins DJ, Staynov DZ, Lee TH: Transcriptional regulation of IL-5 gene expression in human T-cells. J Allergy Clin Immunol 1994;93:679.

72   Akira S, Isshiki H, Sugita T, Tanabe O, Kinoshita S, Nishio Y, Nakajima T, Hirano T, Kishimoto T: A nuclear factor for IL-6 expression (NF-IL-6) is a member of a C/EBP family. EMBO J 1990; 9:1897–1906.

73   Kunsch C, Lang RK, Rosen CA, Shannon MF: Synergistic transcriptional activation of the IL-8 gene by NF-κB p65 (RelA) and NF-IL-6. J Immunol 1994;153:153–164.

74   Romano-Spica V, Georgiou P, Suzuki H, Papas TS, Bhat NK: Role of ETS1 in IL-2 gene expression. J Immunol 1995;154:2724–2732.

75   Plaut M, Pierce JH, Watson CJ, Hanley-Hyde J, Nordan RP, Paul WE: Mast cell lines produce lymphokines in response to cross-linkage of FceRI or to calcium ionophores. Nature 1989;339: 64–67.

76   Henkel G, Weiss DL, McCoy R, Deloughery T, Tara D, Brown MA: A DNase I hypersensitive site in the second intron of the murine IL-4 gene defines a mast cell-specific enhancer. J Immunol 1992;149:3239–3246.

77   Prieschl EE, Gouilleux-Gruart V, Walker C, Harrer NE, Baumruker T: A nuclear factor of activated T cell-like transcription factor in mast cells is involved in IL-5 gene regulation after IgE plus antigen stimulation. J Immunol 1995;154:6112–6119.

78   Kita H, Ohnishe T, Okubo Y, Weiler D, Abrams JS, Gleich GJ: GM-CSF and IL-3 release from human peripheral blood eosinophils and neutrophils. J Exp Med 1991;174:745–748.

79   Moqbel R, Hamid Q, Ying S, Barkans J, Hartnell A, Tsicopoulos A, Wardlaw AJ, Kay AB: Expression of mRNA and immunoreactivity for the granulocyte-macrophage colony-stimulating factor in activated human eosinophils. J Exp Med 1991;174:749–752.

80   Desreumaux P, Janinn A, Colombel JF, Prin L, Pumas J, Emilie D, Torpier G, Capron A, Capron M: IL-5 mRNA expression by eosinophils in the intestinal mucosa of patients with coeliac disease. J Exp Med 1992;175:293–296.

81   Moqbel R, Ying S, Barkans J, Wakelin M, Kimmitt P, Corrigan CJ, Durham SR, Kay AB: Eosinophils transcribe and translate mRNA for IL-4. J Allergy Clin Immunol 1995;95:221.

82   Churchill L, Friedman B, Schleimer RP, Proud D: Production of granulocyte-macrophage colony-stimulating factor by cultured human tracheal epithelial cells. Immunology 1992;75:189–195.

83   Northrop JP, Ho SN, Chen L, Thomas DJ, Timmerman LA, Nolan GP, Admon A, Crabtree GR: NF-AT components define a family of transcription factors targeted in T-cell activation. Nature 1994;369:497–502.

84   Miyatake S, Otsuka T, Yokota T, Lee F, Arai K: Structure of the chromosomal gene for granulocyte-macrophage colony-stimulating factor: Comparison of the mouse and human genes. EMBO J 1985; 4:2561–2568.

85   Arai N, Nomura D, Villaret D, DeWall Malefijt R, Seiki M, Yoshida M, Minoshima S, Fukuyama R, Maekawa M, Kudoh J, Shimizu N, Yokota K, Abe E, Yokota T, Takebe Y, Arai K: Complete nucleotide sequence of the chromosomal gene for human IL-4 and its expression. J Immunol 1989; 142:274–282.

86   Tanabe T, Konishi M, Mizuta T, Noma T, Honjo T: Molecular cloning and structure of the human interleukin-5 gene. J Biol Chem 1987;262:16580–16584.

Prof. T.H. Lee, Department of Allergy and Respiratory Medicine,
4th Floor Hunt's House, Guy's Hospital, London SE1 9RT (UK)

Hall, IP (ed): Genetics of Asthma and Atopy.
Monogr Allergy. Basel, Karger, 1996, vol 33, pp 153–168

....

β₂-Adrenoceptor Polymorphisms and Asthma

I. P. Hall

Department of Therapeutics, University Hospital, Queen's Medical Centre,
Nottingham, UK

Introduction

β₂-Adrenoceptor agonists are the most important group of bronchodilator drugs used in the treatment of asthma [1]. These agents, together with circulating catecholamines, mediate their effect upon the airways by activating the β₂-adrenoceptor, which is present upon a range of important tissues in the lung including airway smooth muscle, neutrophils, eosinophils, alveolar macrophages and airway epithelial cells [2–6]. β₂-Adrenoceptor agonists have two important effects on the airways: firstly these agents are potent bronchodilator drugs, this action being mediated by the relaxation of airway smooth muscle cells [1], and secondly, β₂-adrenoceptor agonists protect against bronchoconstrictor challenge [1]. The precise mechanism whereby this effect occurs is unclear although it may be related to a reduction in inflammatory mediator release in the airways in addition to a direct effect upon airway smooth muscle. This protective effect is irrespective of the challenge employed.

The effects of β₂-adrenoceptor agonists at the cellular level have been extensively studied over recent years. Following binding of the agonist to the β₂-adrenoceptor, the associated G protein (Gs) dissociates and allows Gs_α to bind to adenylyl cyclase increasing the activity of the enzyme, which results in an increase in intracellular cyclic AMP levels [7]. The vast majority of cellular effects of stimulation with β₂-adrenoceptor agonists are believed to be mediated through activation of protein kinase A, which occurs as a consequence of elevation of cyclic AMP. More recently it has also become apparent that Gs_α can have direct effects upon calcium-activated potassium channels in airway smooth muscle, leading to relaxation through a cyclic AMP-independent mechanism [8, 9]. Table 1 shows some of the potentially important effects of stimulation of β₂-adrenoceptor agonists in airway cells.

Table 1. Possible mechanisms underlying β_2-adrenoceptor-mediated relaxation of airway smooth muscle

---

Activation of K_{ca} channels
Membrane hyperpolarization
Inhibition of spasmogen-induced inositol phospholipid hydrolysis
Increased calcium re-uptake into intracellular stores
Inhibition of inositol trisphosphate-induced calcium release
Inhibition of spasmogen-induced calcium influx
Phosphorylation of myosin light chain kinase

The β_2-adrenoceptor was one of the earliest G-protein receptors to be cloned and sequenced [10], and hence the regulation of this receptor has been extensively studied. Site-directed mutagenesis, chimaeric receptor techniques, and deletion mutants have been used to determine the regions of the receptor important for coupling and signal transduction [11–13]. The β_2-adrenoceptor, in common with all G-protein-coupled receptors which have been cloned to date, has 7 transmembrane-spanning α-helical domains and shares structural homology with rhodopsin. Agonist binding to the β_2-adrenoceptor is believed to be dependent upon interaction of the ligand with specific residues in helices III, IV and VI, whereas coupling to Gs is dependent upon amphiphilic residues predominantly in the third intracytoplasmic loop of the receptor [14]. Receptor coupling can be modulated by phosphorylation of sites in the third intracytoplasmic loop of the receptor [14]. Receptor coupling can be modulated by phosphorylation of sites in the third intracytoplasmic loop and in the C-terminal intracellular tail of the receptor [13, 14].

Long-term exposure of the β_2-adrenoceptor to agonist results in a loss of responsiveness. The mechanisms underlying the regulation of receptor coupling have been extensively studied using similar techniques to those used to define structure and function relationships [for review, see 15]. There are three main pathways whereby the receptor can become uncoupled from its signal transduction pathway. Firstly, cyclic AMP produced within the cell as a result of receptor stimulation can activate protein kinase A which in turn is able to phosphorylate specific serine and threonine residues in the third intracytoplasmic loop and in the C-terminal tail of the receptor and hence reduce the ability of the receptor to activate adenlylyl cyclase. Receptor phosphorylation can also occur through a second pathway involving one or more of a family of specific receptor kinases termed β-adrenergic receptor kinases. The net effect of receptor phosphorylation is to reduce responsiveness of the receptor to subsequent challenge with agonist. In addition to receptor phosphorylation, receptors can also be internalized and over longer periods of time changes in the rate of receptor breakdown can occur.

In addition to post-transcriptional control, receptor number can also be controlled by alteration in the rate of transcription (and/or translation) of the β_2-adrenoceptor gene. A reduction in the rate of gene transcription would obviously result in a reduction in receptor number assuming the change occurs in isolation. Receptor expression can also be up-regulated by new receptor synthesis. The promoter region of the β_2-adrenoceptor gene contains a number of regulatory elements which may be involved in these responses. Obviously, the net effect of these different responses (both at the transcriptional and at the cell membrane level) will determine the receptor number and the ability of the cell to respond to stimulation with β_2-adrenoceptor agonists.

β_2-Adrenoceptor Dysfunction in Asthma

Ever since the first proposal that asthma may in part be due to defective β_2-adrenoceptor function [16], controversy has existed regarding the evidence for abnormal β_2-adrenoceptor regulation in asthma. Although altered airway β_2-adrenoceptor function has been difficult to demonstrate in mild asthmatics, there are some data to support altered β_2-adrenoceptor function in animal models of asthma [17, 18], and in severe or fatal asthma [19]. Circulating lymphocytes from normal and asthmatic individuals have been used as a surrogate to study β_2-adrenoceptor regulation, and some studies have supported altered regulation of β_2-adrenoceptor number and/or coupling [20, 21]. In addition, some (but not all) studies have shown abnormal responsiveness of airway smooth muscle in organ bath experiments [22, 23] with reduced relaxation to β_2-adrenoceptor agonists in tissue from asthmatics. Finally, β_2-adrenoceptor function has been shown to be abnormal following viral infection, which is known to worsen asthma [24]. The majority of these studies were performed before it became clear that β_2-adrenoceptor polymorphisms exist, and hence the genotype of individuals was not examined. It is tempting to speculate that some of the inconsistency in the published data in this area may be due to the effects of β_2-adrenoceptor genotype upon responses (see below).

Molecular Biology and Pharmacology of β_2-Adrenoceptor Polymorphisms

In 1992 a group in Cincinnati described a number of polymorphisms within the coding sequence of the human β_2-adrenoceptor gene [25]. The β_2-adrenocep-

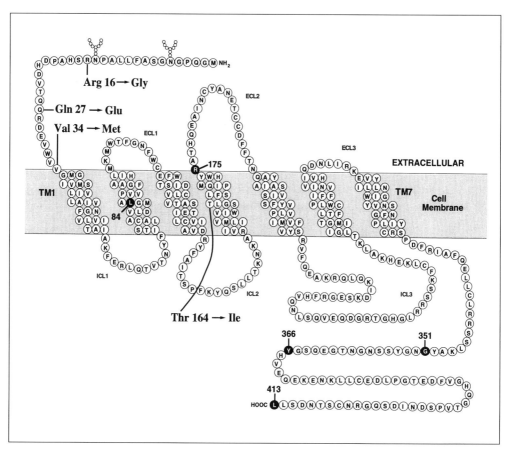

Fig. 1. Predicted structure of the β₂-adrenoceptor showing identified polymorphisms. Note the seven transmembrane-spanning domains (TM1–7) typical of all G-protein-coupled receptors with three intracytoplasmic loops (ICL1–3) and three extracellular loops (ECL1–3). Amino acid residues whose codons contain degenerate polymorphisms are shown in black (84, 175, 351, 366, 413) whereas the identified amino acid polymorphisms are labelled. [Modified from 25, with permission.]

tor is the product of a 1.2-kb intronless gene located on chromosome 5q.31 [10]. Using PCR and temperature gradient gel electrophoresis-based techniques, this group screened genomic DNA from normal and asthmatic patients for possible polymorphisms and proceeded to sequence PCR products with putative base alterations. This study identified a number of polymorphisms (fig. 1), all of which resulted from single base changes at varying positions within the coding region of the gene [25]. Several polymorphisms were degenerate in nature, i.e. they did not result in a change in the amino acid code of the resultant receptor protein because of the redundancy present in the genetic code. However, four polymorphisms were identified which resulted from single base changes and which altered the amino acid sequence of the resultant receptor. Three of these polymorphisms have subsequently been studied in some depth in both transfected and nontransfected cell systems and more recently clinical data have become available indicating their potential relevance in asthma. The fourth polymorphism whereby a methionine substitutes for valine at amino acid 34 (Val-Met 34, fig. 1) did not appear to alter the functional properties of the receptor in initial studies in a transfected cell system and hence has not been studied in nontransformed cells or in human airways.

Arg-Gly 16

A change at base 46 from C to G results in the amino acid sequence of the β_2-adrenoceptor containing a glycine at position 16 instead of an arginine. Amino acid 16 is adjacent to a glycosylation site, the N-aspartamine residue at amino acid 15, and hence although at first sight this N-terminal portion of the receptor may seem an unlikely site for a physiologically relevant polymorphism the Arg-Gly 16 substitution could potentially alter the way in which the receptor is processed by the cell. There are some data to support the contention that the N-terminal region of G-protein-coupled receptors is important in insertion of the receptor into the cell membrane [26–29]: for example, N-terminal deletions of the β_2-adrenoceptor are processed abnormally when expressed in transfected cell systems. Similar alterations in receptor processing can be seen with N-terminal abnormalities in other G-protein-coupled receptors, for example one form of retinitis pigmentosa is due to an N-terminal deletion in the rhodopsin gene [27].

With regard to the β_2-adrenoceptor, evidence for the structural relevance of the amino acid 16 substitution came initially from work where the polymorphism was reproduced by site-directed mutagenesis. The Gly 16 form of the receptor was transfected into CHW cells which do not constitutively express the β_2-adrenoceptor. When the transfected receptor was studied in this system, ligand binding and receptor coupling remained unaltered upon initial contact with an agonist [26]. However, prolonged contact with agonist resulted in markedly enhanced down-regulation of the receptor which was manifested by a reduced responsiveness of the cell at the level of cyclic AMP accumulation and a reduction in receptor number assessed by ligand-binding studies [26].

In subsequent studies we have examined the functional relevance of the Gly 16 polymorphism in human primary airway smooth muscle cell cultures [28]. These cells are derived from human trachealis and express the β_2-adreno-

ceptor constitutively. We have previously studied β_2-adrenoceptor characteristics in these cells extensively, and shown that the regulation of adenylyl
cyclase coupling is very similar to that seen in tissue obtained ex vivo. The cells
are nontransformed but can be grown over repeated passage before eventually
becoming senescent: receptor expression remains constant over this period
[30]. The big advantage of this system is that receptor expression is regulated
by the 'native' regulatory elements, in contrast to the transfected cell system,
where expression is dependent upon regulation by foreign regulatory sequences,
usually derived from viral vectors which may behave quite differently from
the constitutive human regulatory elements. In addition, the level of receptor
expression is likely to be more physiological. This is particularly important
when down-regulation is being studied, because down-regulation will be much
more readily demonstrable in a transfected cell system wherein receptors are
overexpressed. We have derived cell populations from a number of individuals
without asthma, and then genotyped these populations with regard to their
β_2-adrenoceptor polymorphisms. Cells homozygous for Gly 16 were studied,
and in these experiments we found that prolonged exposure to agonist resulted
in a marked reduction in the cyclic AMP response to rechallenge. Because of
the profound degree of desensitization seen in these experiments, no marked
difference was observed when cyclic AMP responses were compared in cells
homozygous for the Arg 16 and Gly 16 forms of the receptor. However,
a significantly greater reduction was observed in receptor number in cells
homozygous for Gly 16 following prolonged exposure to agonist suggesting
that the Gly 16 form of the receptor when constitutively expressed does down-
regulate to a greater extent than the Arg 16 form of the receptor [28] (table
2). Although the Arg 16 form of the receptor is traditionally designated as
'wild type' then the Gly 16 form of the receptor is actually more frequent and
should perhaps be considered as the true wild type. The allelic frequency of
these forms of the receptor in the Nottingham population is Arg 16, 33% and
Gly 16, 67% [unpubl. data].

Gln–Glu 27

A change from a C to G at base number 76 results in Glu 27 being altered
to glutamic acid. This polymorphism has been studied using the same techniques
as those described above for the Arg-Gly 16 polymorphism. Agonist binding,
antagonist binding and agonist-promoted adenylyl cyclase activation were unaltered when the Glu 27 form of the receptor was studied in the transfected CHW
cell system [26]. However, in marked contrast to Gly 16, Glu 27 demonstrates
markedly reduced down-regulation following prolonged agonist exposure in

Table 2. Effect of β_2-adrenoceptor polymorphisms on β_2-adrenoceptor down-regulation in cultured human airway smooth muscle cells

β_2-Adrenergic receptor genotype	B_{max} baseline fmol/mg	B_{max} down-regulation fmol/mg	Down-regulation % (of baseline binding)	Desensitization % (ISO, 1 μM) (cAMP)
Wild type	86.5 ± 16.1	20.6 ± 9.0	77.8 ± 8.1	90 ± 5
Arg 16→Gly	65.3 ± 15.3	$9.7 \pm 6.6*$	$95.6 \pm 1.7*$	84 ± 5
Gln 27→Glu	63.7 ± 12.7	$46.2 \pm 15.9*$	$29.5 \pm 12.7*$	$33 \pm 7*$

Characteristics of β_2-adrenoceptor down-regulation in cultured human airway smooth muscle cells following exposure to isoprenaline. Primary cultures of human airway smooth muscle cells homozygous for the β_2-adrenoceptor polymorphisms were studied. Levels of baseline (unstimulated) β_2-adrenoceptor binding and binding following exposure to 1 μM isoprenaline overnight are shown in absolute terms and as a percentage of baseline binding. Desensitization of the cyclic AMP response to a subsequent challenge with 1 μM isoprenaline (ISO) for 10 min is also shown as a percentage of the response in naive cells. [Adapted from 28.]
*p<0.01 cf. wild type.

both transfected and nontransfected cell systems homozygous for the polymorphism. In primary cultures of human airway smooth muscle cells, when assessed at the level of cyclic AMP formation, the Glu 27 form of the receptor results in approximately a 60-fold shift to the right in the dose-response curve for desensitization to isoprenaline [28]. Resistance to down-regulation was also observed in radioligand-binding studies (table 2) [28].

There are a number of possible sites for altered regulation of the β_2-adrenoceptor which might account for the different down-regulation profiles of the Gly 16 and Glu 27 forms of the receptor. The expression of the β_2-adrenoceptor on the cell is obviously a dynamic process, the exact level of expression being dependent upon the balance between the rate of new receptor synthesis and the rate of receptor loss. Liggett's group have attempted to define the site of altered regulation in a systematic manner. Initial studies were designed to measure the rate of receptor protein synthesis by irreversibly alkylating receptors with the modified β-adrenergic antagonist BIM (alkylating pindolol). These studies found no differences in the rate of new receptor synthesis when the different polymorphic forms of the receptor were studied. To examine the effects of the polymorphisms on cell trafficking of receptors, β_2-adrenoceptor density was assessed for each form of the receptor in fractions of cell lysates separated on sucrose density gradients. Again, no marked differences were observed in the lysates from CHW cells transfected with the different forms of the receptor; these results were supported by binding studies using the hydrophilic ligand [^{3}H]CGP 12177 in whole cell binding studies. Transcription of the different forms of the receptor was not significantly altered following agonist exposure, although there was a small increase (nonsignificant) seen in the mRNA levels of the Glu 27 variant. However, the magnitude of this effect, coupled with the observation that a significant decrease in mRNA levels for the 'wild-type' form of the receptor did not occur following agonist exposure, suggests that this effect does not account for the observed differences in receptor down-regulation. Instead, it seems more likely that altered receptor degradation may be important. Using a polyclonal antisera directed against the C-terminal region of the receptor, the Glu 27 form of the receptor was found to display altered mobility in gel electrophoresis assays, possibly due to an alteration in the glycosylation state of the receptor. Both the Gln 27 and Glu 27 forms of the receptor are common in the general population, with allelic frequencies of 55 and 45% respectively.

Thr-Ile 164

A base change from C to a T in codon 164 results in Thr 164 becoming isoleucine (Thr-Ile 164). This polymorphism is much rarer than the amino acid 16 and 27 polymorphisms, with an allelic frequency of around 1–3% in the US population [25]. Hence homozygous individuals are extremely uncommon and no data are available in nontransfected cells to date. However in a transfected cell system where the Ile 164 form of the receptor was expressed, the ligand-binding properties were markedly shifted [31]. This form of the receptor binds agonists less avidly by a factor of approximately 4-fold compared with the wild-type (Thr 164) form of the receptor. The mechanism underlying this altered agonist affinity is believed to be due to the altered environment of the ligand-binding site in the β$_2$-adrenoceptor. Previous studies have suggested that a serine at amino acid 165 is capable of forming a hydrogen bond with the β-hydroxyl group of catecholamine ligands [12]. The lower affinity of the Ile 164 form of the receptor for ligands may be due to an alteration in the charge distribution at this position in the fourth transmembrane-spanning domain. This appears to be a functionally relevant alteration in binding affinity because the ability of the Ile 164 from of the receptor to stimulate adenylyl cyclase in transfected CHW cells was markedly reduced. Further evidence for this hypothesis comes from the observation that the ability of ligands without this β-hydroxyl group to stimulate the receptor is preserved [31]. Radioligand-binding studies have also suggested that the ability of the Ile 164 form of the receptor to form high-affinity ternary complexes (necessary for efficient coupling) is reduced, which may also contribute to the reduced ability of this form of the receptor to stimulate adenylyl cyclase. However, the true physiological relevance of these observations can only be confirmed by studying the Ile 164 form of the receptor when constitutively expressed which, given its rarity, may prove difficult.

Restriction Fragment Length Polymorphisms (RFLP) and the β₂-Adrenoceptor

In addition to the above polymorphisms, a Japanese group has described a Ban 1 RFLP linked to the human β₂-adrenoceptor in the Japanese population [32]. However, the exact site of the mutation causing this RFLP remains to be determined and it may even prove to be outside the coding region of the gene. At present it is unclear how this RFLP relates to the known polymorphisms described above. One possibility is that the RFLP may occur due to the degenerate polymorphism previously identified by Liggett's group at base number 523 [25] although if this is the case it is difficult to believe that the receptor will behave markedly differently because the presence of a substitution at this base does not alter the amino acid code of the receptor. However, it is possible that this polymorphism is in linkage disequilibrium with one of the other functionally relevant β₂-adrenoceptor polymorphisms.

Association Studies in Asthma

In view of the demonstration in linkage studies that markers for surrogates for the asthma phenotype map to the 5q region containing the gene for the β₂-adrenoceptor [33,34] and the observation that the different polymorphic forms of the β₂-adrenoceptor confer physiologically relevant alterations in the way the receptor behaves, the relevance of these β₂-receptor polymorphisms to airway function has been the subject of preliminary studies in asthmatic and nonasthmatic individuals. Before describing these studies, two important points must be borne in mind. Firstly all individuals will have two genes for the β₂-adrenoceptor and therefore can potentially be homozygous or heterozygous for any of the given polymorphisms. We have also observed that linkage disequilbrium occurs between the Gly 16 and Glu 27 forms of the receptor [unpubl. data]. In view of the proximity of the β₂-adrenoceptor to other candidate genes for bronchial hyperreactivity (most notably the cluster of Th₂ cytokine genes) and the asthma phenotype it should also be remembered that associations between β₂-adrenoceptor polymorphisms and clinical markers of asthma may also arise due to linkage disequilibrium of the polymorphisms with other possible loci in the same region of chromosome 5q. However, despite these reservations interesting preliminary data exist implicating the relevance of β_2-adrenoceptor polymorphisms in asthma.

Clinical Studies

Following the initial description of polymorphisms within the β_2-adrenoceptor, a search was made for associations between polymorphisms and markers of asthma severity. The initial study used clinical markers of asthma severity based upon patient history and medication history, and examined the distribution of the β_2-adrenoceptor polymorphisms in 51 patients with moderately severe asthma and 56 normal subjects [25]. In this study no polymorphism was significantly commoner in asthmatics compared with normal subjects, although the power of this study to identify a weak association with asthma was low. However, when severity was gauged on the basis of medication requirements, there was a significant association between the requirement for continuous oral steroids and the homozygous Gly 16 genotype. Following this initial observation, and given the in vitro data demonstrating alteration in the down-regulation characteristics of the β_2-adrenoceptor dependent upon genotype, further studies have been performed to investigate potential relationships between β_2-adrenoceptor genotype and markers of asthma severity.

Gly 16 and Nocturnal Asthma

Given that the Gly 16 form of the β_2-adrenoceptor down-regulates to a greater extent than the Arg 16 form of the receptor in vitro, the Cincinnati group set out to study the hypothesis that individuals with severe nocturnal worsening of their asthma may more commonly have this form of the receptor. Previous work has demonstrated that subjects with marked nocturnal worsening of their asthma (defined as a > 10% fall in FEV_1 at 04:00 vs. 16:00 h) have a greater fall in β_2-adrenoceptor density on circulating mononuclear leucocytes [35]. Twenty-three subjects with documented nocturnal asthma (mean % decrease in peak expiratory flow rate 34%) were compared with 22 subjects with asthma who did not demonstrate a nocturnal decline in peak expiratory flow rate (mean % decrease 2.3%). These subjects were then genotyped for polymorphisms at loci 16, 27 and 164. Subjects with the Gly 16 form of the receptor were 6 times more likely to have nocturnal asthma [36]. No such relationship was observed for locus 27 or 164 polymorphisms, although the allelic frequency of the 164 polymorphism is too low for a study of this size to assess its potential contribution.

Gly 16 and Increased Airway Reactivity

Preliminary results from a large Dutch study of asthmatic families have suggested an association between airway hyperreactivity (defined as histamine $PD_{20} < 8$ µmol) and the Gly 16 genotype [37]. In this same population, linkage was also observed for markers close to the β_2-adrenoceptor locus on 5q31 and airway reactivity [34].

Glu 27 and Airway Reactivity

A recent study from our group has examined the possible relationship between the Glu 27 genotype and airway reactivity in mild to moderate asthmatic subjects [38]. Our hypothesis was that individuals with the form of the β_2-adrenoceptor which down-regulates to a lesser extent might have less severe airway inflammation. To test this hypothesis we studied 65 subjects with mild to moderate asthma, in all of whom a methacholine PD_{20} measurement could be obtained. All subjects were on an inhaled steroid and 'as required' inhaled β-agonist medication at the time of assessment In this study we found that asthmatic subjects who were homozygous for the Glu 27 form of the β_2-adrenoceptor were approximately 4-fold less reactive to inhaled methacholine than subjects who were homozygous for the Gln 27 form of the receptor, with the group of heterozygous individuals having an intermediate mean PD_{20} value (fig. 2) [38]. The groups were well matched in terms of other baseline characteristics, although subjects with Glu 27 were slightly more likely to be atopic (which would be expected to lessen the association) and also had a higher starting FEV_1. However, if Gln 27 is associated with increased airway reactivity these subjects might be expected to have more severe asthma and hence have a lower starting FEV_1.

Although this study demonstrates an association between Glu 27 form of the β_2-adrenoceptor and lower airway reactivity, further studies are required to assess the potential contribution of this polymorphism to airway responses. Taken to its logical conclusion, one would predict that the Glu 27 genotype may in some individuals protect against the development of asthma. However, even if this polymorphism only contributed 10% to the risk of developing asthma, much larger studies than those which have been performed to date would be required to assess the relationship. These studies are currently in progress and should clarify the contribution of both the loci 16 and 27 polymorphisms to the asthma phenotype and to asthma severity. One would also predict that if these associations are real, then linkage studies should demonstrate an association between the asthma phenotype, or markers of asthma severity such as airway reactivity, and the region on chromosome 5q31 where

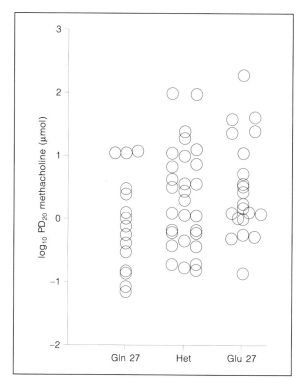

Fig. 2. Methacholine PD$_{20}$ values in mild to moderate asthmatic individuals genotyped for the β$_2$-adrenoceptor amino acid 27 polymorphism. log$_{10}$ methacholine PD$_{20}$ values are shown for individuals homozygous for Gln 27, individuals homozygous for Glu 27 and for heterozygous individuals. The mean methacholine PD$_{20}$ values (µmol) and confidence intervals for the three groups were as follows: Gln/Gln 27, 0.86 (0.25–2.15); Gln/Glu 27, 1.96 (1.08–5.01); Glu/Glu, 27, 3.23 (1.25–7.40) (p=0.034 Glu 27 versus Gln 27 homozygotes). [From 38, with permission.]

the β$_2$-adrenoceptor is sited. A number of studies have suggested linkage of asthma or bronchial hyperreactivity with markers on 5q close to the β$_2$-adrenoceptor locus [33, 34], although it must be remembered that the major cytokine gene cluster for Th$_2$ cytokines lies nearby around 5q23–31 [39].

Concluding Remarks

Whereas the vast majority of work on the genetics of asthma has used standard linkage approaches to define candidate loci, the potential contribution of β$_2$-adrenoceptor polymorphisms has used exactly the opposite approach, namely the candidate gene approach. It is clear from both work in transfected cell lines and also in primary cultures of airway cells that β_2-adrenoceptor polymorphisms can modulate the behaviour of the β_2-adrenoceptor by altering the characteristics of down-regulation of the receptor following prolonged agonist exposure. The precise cellular mechanisms underlying these effects remain to be determined. Because the amino acid 16 and 27 polymorphisms are relatively abundant in the general population, these polymorphisms could potentially be an important cause of biological variability between individuals. The preliminary data described in this chapter suggests that the Gly 16 genotype is associated with increased airway reactivity and more severe asthma and in particular with nocturnal worsening of asthma. In contrast, the Glu 27 genotype is associated with less reactive airways in asthmatic subjects. Whether either of these polymorphisms alter an individual's susceptibility to develop asthma per se remains to be established. Because both of these polymorphisms are common then some individuals will be homozygous for both Gly 16 and Glu 27 polymorphisms, although these appear in the initial studies to be in linkage disequilibrium [Wheatley et al., unpubl. data]. The effect of having these polymorphisms in combination and the phenotype of heterozygous individuals remains to be fully established. In addition to the airway studies, there have been preliminary studies assessing the possible relevance of β_2-adrenoceptor polymorphism in cardiovascular disease. There is marked variation in the vascular response to infused isoprenaline in different populations [39, 40], and differences have also been seen in β_2-adrenoceptor density in cultured skin fibroblasts in salt sensitive and insensitive individuals [41]. A recent abstract has reported that the β_2-adrenoceptor RFLP described above is associated with hypertension in blacks [42], and studies are currently underway examining whether vascular responses to infused β_2-agonists may be in part genetically determined.

Finally the message that common polymorphisms exist within important airway receptor genes and that these polymorphisms have physiologically relevant effects upon receptor functioning is of broader significance to airway research. To date, few searches have been made for polymorphisms in other airway receptors and this is a potentially fruitful area for future research.

Acknowledgements

Some of the work described in this chapter was funded by the NAC and the MRC (UK). The author would like to thank S. Liggett for permission to use Figure 1 and S. Spiller for secretarial assistance.

References

1   Hall IP, Tattersfield A: Beta agonists; in Clark TJ, Godfrey S, Lee T (eds): Asthma. London, Chapman & Hall, 1992, pp 341–365.
2   Carstairs JR, Nimmo A, Barnes P: Autoradiographic visualization of β-adrenoceptor subtypes in human lung. Am Rev Respir Dis 1985;132:541–547.
3   Hamid QA, Mak J, Sheppard M, Corrin B, Venter J, Barnes P: Localization of beta-2-adrenoceptor messenger RNA in human rat lung using in situ hybridization: Correlation with receptor autoradiography. Eur J Pharmacol 1991;206:133–138.
4   Liggett SB: Identification and characterisation of a homogenous population of β_2-adrenergic receptors on human alveolar macrophages. Am Rev Respir Dis 1989;139:552–555.
5   Galant SP, Duriseti L, Underwood S, Allred S, Insel P: Beta-adrenergic receptors of polymorphonuclear particulates in bronchial asthma. J Clin Invest 1980;65:577–585.
6   Yukawa T, Ukena D, Kroegel C, Chanez P, Dent G, Chung K, Barnes P: Beta-adrenergic receptors on eosinophils. Binding and functional studies. Am Rev Respir Dis 1990;141:1446–1452.
7   Liggett SB, Raymond J: Pharmacology and molecular biology of adrenergic receptors; in Bouloux PM (ed): Catecholamines. Baillière's Clinical Endocrinology and Metabolism, ed 7. London, Saunders, 1993, pp 279–306.
8   Kume H, Graziano MP, Kotlikoff MI: Stimulatory and inhibitory regulation of calcium-activated potassium channels by granine nucleotide-binding proteins. Proc Natl Acad Sci 1992;89:11051–11055.
9   Kume H, Hall I, Washabau R, Takagi K, Kotlikoff M: Beta-adrenergic agonists regulate K_{ca} channels in airway smooth muscle by cAMP dependent and independent mechanisms. J Clin Invest 1994; 93:371–379.
10 Kobilka BK, Dixon R, Frielle H, Dohlam M, Bolanowski I, Sigal I, Yang-Fen T, Francke U, Caron M, Lefkowitz R: cDNA for the human β_2-adrenergic receptor: A protein with multiple membrane spanning domains and encoded by a gene whose chromosomal location is shared with that of the receptor for platelet-derived growth factor. Proc Natl Acad Sci USA 1987;84:46–50.
11 Liggett SB, Freedman N, Schwinn D, Lefkowitz R: Structural basis for receptor subtype specific regulation revealed by a chimeric β_1-β_2 adrenergic receptor. Proc Natl Acad Sci USA 1993;90:3665-3669.
12 Strader CD, Candelore, M, Hill W, Sigal I, Dixon R: Identification of two serene residues involved in agonist activation of the β-adrenergic receptor. J Biol Chem 1989;264:3572–3578.
13 Liggett SB, Lefkowitz R: Adrenergic receptor-coupled adenylyl cyclase systems: Regulation of receptor function by phosphorylation, sequestration and down-regulation; in Sibley D, Houslay M, (eds): Regulation of Cellular Signal Transduction Pathways by Desensitization and Amplification. London, Wiley, 1993, pp 71–97.
14 Strader CD, Signal U, Dixon A: Mapping the functional domains of the beta-adrenergic receptor. Am J Respir Cell Mol Biol 1989;1;81–86.
15 Hausdorff WP, Caron M, Lefkowitz R: Turning off the signal: Desensitization of beta-adrenergic receptor function. FASEB J 1990;4:2881–2889.
16 Szebtivanyl A: The beta-adrenergic theory of the atopic abnormality in bronchial asthma. J Allergy 1968;42:203–232.
17 Barnes P, Dollery C, MacDermot J: Increased pulmonary α-adrenergic and reduced β-adrenergic receptors in experimental asthama. Nature 1980;285:569–571.
18 Tobias D, Sauder R, Hirschman C: Reduced sensitivity of beta-adrenergic agonists in basenji-greyhound dogs. J Appl Physiol 1990;69:1212–1219.
19 Bai TR: Abnormalities in airway smooth muscle in fatal asthma. Am Rev Respir Dis 1991;143: 441–443.
20 Kariman K: β-Adrenergic receptor binding in lymphocytes from patients with asthma. Lung 1980; 158:41–51.
21 Conolly ME, Greenacre J: The lymphocyte β-adrenoreceptor in normal subjects and patients with bronchial asthma. J Clin Invest 1976;58:1307–1316.
22 Goldie RG, Spina D, Henry P, Lulich K, Paterson J: In vitro responsiveness of human asthmatic bronchus to carbachol, histamine, β-adrenoceptor agonists and theophylline. Br J Clin Pharmacol 1986;22:669-676.

23 Whicker SD, Armour C, Black J: Responsiveness of bronchial smooth muscle from asthmatic patients to relaxant and contractile agonists. Pulm Pharmacol 1988;1:25–31.

24 Buckner CK, Clayton D, Ain-Shoka A, Busse W, Dick E, Shult P: Parainfluenza 3 infection blocks the ability of a beta-adrenergic receptor agonist to inhibit antigen-induced contraction of guinea pig isolated airway smooth muscle. J Clin Invest 1981;67:376–384.

25 Reihaus E, Innis M, MacIntyre N, Liggett S: Mutations in the gene encoding for the β_2-adrenergic receptor in normal and asthmatic subjects. Am J Respir Cell Mol Biol 1993;8:334–339.

26 Green SA, Turki J, Innis M, Liggett S: Amino-terminal polymorphisms of the human β_2-adrenergic receptor impart distinct agonist-promoted regulatory properties. Biochemistry 1994;33:9414–9419.

27 Raymond JR: Hereditary and acquired defects in signaling through the hormone receptor G protein complex. Am J Physiol 1994;266:F163–F174.

28 Green SA, Turki J, Bejarano P, Hall I, Liggett S: Influence of β_2-adrenergic receptor genotypes on signal transduction in human airway smooth muscle cells. Am J Respir Cell Mol Biol 1995;13: 25–33.

29 Easton MG, Jacinto M, Theiss C, Liggett S: The palmitoylated cysteine of the cytoplasmic tail of α_{2A}-adrenergic receptors confers subtype-specific agonist-promoted down-regulation. Proc Natl Acad Sci USA 1994;91:11178–11182.

30 Hall IP, Kotlikoff M: Use of cultured airway myocytes for study of airway muscle. Am J Physiol 1995;268:L1–L11.

31 Green SA, Cole G, Jacinto M, Innis M, Liggett S: A polymorphism of the human β_2-adrenergic receptor with the fourth transmembrane domain alters ligand binding and functional properties of the receptor. J Biol Chem 1993;268:23116–23121.

32 Ohe M, Munakata M, Hizawa N, Itoh A, Doi I, Yamaguchi E, Homma Y, Kawakami Y: Beta-2-adrenergic receptor gene restriction fragment length polymorphisms and bronchial asthma. Thorax 1995;50:353–359.

33 Marsh DG, Neely J, Breazeale D, Ghosh B, Friedhoff L, Ehrlich-Kautzy E, Krishnawamy G, Beaty T: Linkage analysis of IL-4 and other chromosome 5q31.1 markers and total serum immunoglobulin E concentrations. Science 1994;264:1152–1155.

34 Postma DS, Bleecker E, Amelung P, Holroyd K, Xu J, Panhuysen C, Meyers D, Levitt R: Genetic susceptibility to asthma-bronchial hyperresponsiveness co-inherited with a major gene for atopy. N Engl J Med 1995;333:894–900.

35 Szefler SJ, Ando R, Cicutto C, Surs W, Hill M, Martin R: Plasma histamine, epinephrine, cortisol and leukocyte β-adrenergic receptors in nocturnal asthma. Clin Pharmacol Ther 1991;49:59–68.

36 Turki J, Pak J, Green S, Martin R, Liggett S: Genetic polymorphisms of the β_2-adrenergic receptor in nocturnal and non-nocturnal asthma: Evidence that Gly 16 correlates with the nocturnal phenotype. J Clin Invest 1995;95:1635–1641.

37 Holroyd KJ, Levitt R, Dragwa C, Amelung P, Panhuysen C, Meyers D, Bleecker E, Postma D: Evidence for β_2-adrenergic receptor polymorphism at amino acid 16 as a risk factor for bronchial hyperresponsiveness. Am J Respir Crit Care Med 1995;151:A673.

38 Hall IP, Wheatley A, Wilding P, Liggett S: Association of the Glu 27β_2-adrenoceptor polymorphism with lower airway reactivity in asthmatic subjects. Lancet 1995;345:1213–1214.

39 Eichler HG, Blaschke T, Hoffman B: Decreased responsiveness of superficial hand veins to phenylephrine in black normotensive males. J Cardiovasc Pharmacol 1990;16:177–181.

40 Lang CC, Stein C, Brown R, Deegan R, Nelson R, Huai B, Wood M, Wood A: Attenuation of isoproterol-mediated vasodilation in blacks. N Engl J Med 1995;333:155-160.

41 Kotanko P, Hoglinger O, Skrabal F: β_2-adrenoceptor density in fibroblast culture correlates with human NaCl sensitivity. Am J Physiol 1992;263:C623–C627.

42 Svetkey LP, Timmons P, Emovon O, Dawson D, Lefkowitz R, Anderson N, Chen Y-T: Association between essential hypertension in blacks and β_2-adrenergic receptor genotype (abstract). Hypertension 1995;26:575.

Dr. I.P. Hall, Department of Therapeutics, University Hospital, Queen's Medical Centre, Nottingham NG7 2UH (UK)

Subject Index

Asthma (continued)
  phenotype 2
  TNF 125–135
Atopic allergy 59–64
Atopic asthmatics 139
Atopic dermatitis 139
Atopy 13–15, 53, 115
  asthma 19
  bronchial hyperresponsiveness 110, 111
  cytokine genes 138–147
  definition 6, 7, 116
  genetics 1, 2, 99–104
  HLA class II genes 65
  IgE 8
  origins 97, 98
Autosomal dominant transmission 112

B cell epitopes 73
B*1/B*1 genotype 134
Basenji-greyhound dog, model for
  bronchial hyperreactivity 36–41, 48
Bet v I 73, 78, 84
Birch tree pollen 78, 84
BN rat strain 43
Breastfeeding, allergic disease 26
Bronchial hyperreactivity 1, 2, 15–18, 115
  asthma 109–111
  definition 6, 7
  dog models 36–41, 48
  rodent models 41–49
Bronchoalveolar lavage, 139
Bronchoconstriction 15, 36
  dogs 37

cAMP 153, 154
Can f I 73
Candidate loci 114, 118
Carbachol 17
  response by rat 43
Carter effect 21
CCAAT element 144
CD3 complex 79, 80
CD3+ T cells 139
CD4 coreceptor 79
CD8 coreceptor 79
CD20 101
Cedar pollen 78
C3H/HeJ strain of mouse 45, 46, 48
Chinese descent, IgE 13

Cholinergic agonists 37
  rat 43
Chromosome 5 2, 3, 118, 119
Chromosome 11 3
Chromosome 11q, atopy 100, 101
Chromosome 11q13 10, 120
Cigarette smoke, allergic disease 24
Citric acid 44
CKE2 motif 141
Class I MHC 74, 75
Class II MHC 75
CLE0 motif 141, 142
CLE1 motif 141
Corticosteroids 139
Cytokine-1 140, 141
Cytokine-2 141
Cytokines 3
  asthma 114
  regulation 138–147

D11A534 115–117
D11S480 115, 116
D11S527 115–117
Der p I 63, 66, 73, 79
Der p II 73, 79, 87
Diet, allergic disease 25, 26
Dizygotic twins, asthma 21, 22
DNA
  markers 113, 114
  typing, HLA class II molecules 62
Dogs, bronchial hyperreactivity 36–41, 48
Double antibody radioimmunoassay 11
Down regulation of receptors 158, 159, 162,
  163, 165
DP subregion of HLA class II molecule 55
DPA1 76
DPA2 76
DPB1 76
DPB2 76
DQ subregion, HLA class II molecule 55
DQA1 76
DQA2 76
DQB1 76
DQB2 76
DR subregion, HLA class II molecule 55
DRB1*08 65
DRB1*15 78
DRB3*0101 78
DRB4 65

Ii chain 57, 58
IL-2 147
IL-4 114, 118, 119
  atopic disease 138–140
  regulation 140, 143, 144, 146, 147
IL-5 144
  atopic disease 138–140
  regulation 145–147
IL-5A 145
IL-5C 145
IL-5CLE0 145
IL-5P 145
IL-9 118
IL-13 118
Immune disease, HLA 58
Immunoglobulins 71, 72
Inbreeding 42, 44
Indoor air, allergic disease 25
Interferon stimulatory response element 143
γ-Interferon 139
Intradermal skin test score 14
Intrauterine development, allergic
  disease 26
IRF-1 118
IRF-2 143
Isocyanate-induced asthma 64

Leu 181 102–104
Leucine, IgE responsiveness 102, 103
Leukocyte antigen genes 53–67
Leukotriene D4 44
Linkage analysis 111, 112, 114
Linkage disequilibrium 77, 120, 161,
  165
LMP2 57
LMP7 57
Logarithm of odds ratio 47
Lol p I 78, 84
Lol p II 78
Lol p III 78
LT-a 127
  secretion 130
LT-b 127
Lung resistance, antigen challenge 36

m₂ muscarinic receptor subtype 37, 41, 43
Magnesium, allergic disease 26
Major histocompatibility complex 72,
  74–77

Markers 48, 115, 116
Mast cells 101, 102, 146
  allergy 38
Maternal effect on asthma 20, 21
Maternal smoking 24
Methacholine 16, 18, 35, 163, 164
  sensitivity in dogs 37
β-2-Microglobulin 74
Migrant studies 12, 13, 23
Mites, allergy to, see House dust mite allergy
Mold allergen 61
Monozygotic twins, asthma 21–23
Month of birth, allergic disease 27
Mouse
  hyperreactive strains 45–47
  hyporeactive strains 45–47
  model of bronchial hyperreactivity 44–49
Multiple sclerosis 82, 83
Muscarinic receptors 37, 41, 43
Myelin basic protein 82

NcoI RFLP 129–132, 134
Negative regulatory elements 144
Neonatal diet, allergic disease 25, 26
NF-1 143, 144
NF-CLEOa 142
NF-CLEOb 142
NF(P) 143
NF-Y 144
NRE-1 144
NRE-2 144
Nucleotide substitutions 102

Occupation, allergic disease 27
Ovalbumin, responses 43, 44

p55 receptor 127
p75 receptor 127
Palindromic element 142, 143, 147
Parasympathetic nervous system 36
PCR 86
PEBP2 142
Pedigrees, human 10
Peptide binding groove 54, 56
Peptide complex, T cell receptor 59
Peptides
  binding 57, 58
  transporting 57
Pets, household, allergic disease 25